THE M & E HANDBOOK SERIES

Introduction to Ecology

J.C. Emberlin

BSc, PhD
Senior Lecturer in Ecology
Department of Geography,
The Polytechnic of North London

MACDONALD AND EVANS

Macdonald & Evans Ltd
Estover, Plymouth PL6 7PZ

First published 1983
Reprinted in this format 1984

© Macdonald & Evans Ltd 1983

British Library Cataloguing in Publication Data

Emberlin, J.C.
  Introduction to ecology. — (The M & E
  HANDBOOK series)
  1. Ecology
  I. Title
  574.5   QH541

  ISBN 0-7121-0965-X

Phototypeset by *Sunrise Setting*, Torquay, Devon
Printed in Great Britain by
Hollen Street Press Ltd, Slough

Preface

This HANDBOOK gives a concise presentation of the main aspects of ecology, a subject which is becoming increasingly important as an area of study in schools and colleges, both as a part of syllabuses in biology, zoology, botany and geography and as a component of environmental studies, general studies and environmental management courses.

The book is designed primarily for use in preparation for Advanced Level GCE examinations but it caters also for some work at Ordinary Level and forms a useful introduction to ecology for undergraduates and interested lay people.

The text not only covers ecosystem structure and function, productivity, environmental factors, population dynamics and evolution but also examines strategies for life, migration, distribution patterns, major natural ecosystems, British habitats and practical techniques for collecting data in the field. It concludes by focusing on the role of man as an ecological factor and reviews the importance of conservation. Throughout the book emphasis is given to the latest ideas and current research in ecology. Many topics which were thought to be simple a few decades ago are now seen as complex with no simple explanation. In these cases alternative theories are presented and discussed.

Each chapter is followed by a progress test which is planned to assess the readers' comprehension of the material. Recent examination questions from relevant "A" Level and "AO" Level syllabuses are included in Appendix IV at the end of the book. Where possible, the proportions of marks assigned to individual parts of each question have been indicated. However, since ecology has been added to many syllabuses relatively recently, the selection of previous questions from some examination boards is limited. The appendix on examination techniques provides helpful suggestions on methods of answering the various types of question.

As this HANDBOOK is concise candidates should refer to works recommended in the selected reading section (See

Appendix I) to study topics in depth. Ideally ecology should be studied by practical investigation and experimentation. The author advises therefore that, whenever possible, field work should be conducted to study individual habitats using the techniques described in XIII.

Acknowledgments. The author would like to thank David Snushall of the Geography Department, Polytechnic of North-London for drawing the figures in this book. Thanks are also due to Mrs M. Begum for technical assistance, Miss P. Dolphin for help with the reading lists and to Miss Hilary Bryant for valuable general help. Sample examination questions (*see* Appendix IV) are reproduced with the kind permission of the Associated Examining Board, the University of Cambridge Local Examinations Syndicate, the Oxford Delegacy of Local Examinations, the Oxford and Cambridge Schools Examination Board and the University of London University Entrance and School Examinations Council.

Acknowledgment is made for permission to reproduce the following figures. Figs 13 and 14 modified from *Fundamentals of Ecology* (3rd ed) by Eugene P. Odum. Copyright © 1971 by W.B. Saunders Co., reprinted by permission of Holt, Rhinehart and Winston, CBS College Publishing; Fig. 15 after M. Ashby, *Introduction to Plant Ecology*, Macmillan, London and Basingstoke 1968; Fig. 17 after G.M. Woodwell, "The Ecological Effects of Radiation", *Sci. Am.* 208 (6), 1963; Fig. 18 after A.G. Tansley, *The British Isles and their vegetation*, Cambridge University Press 1939; Fig. 24 after R.C. Ward, *Principles of Hydrology*, McGraw-Hill 1967; Fig. 25 after C.W. Thornthwaite, "An approach towards a rational classification of climate" *Geog. Rev.* (38) 1948; Fig. 41 (after Fig. 6.3 p.120 *Biogeography: A study of plants in the ecosphere*, J. Tivy. Second edition published by Longman Group Ltd, 1982. First published by Oliver Boyd Ltd, 1971.) using data compiled from R. Good, *The Geography of flowering plants* (2nd ed) Longman Group Ltd., 1953; Fig. 56 after P. Danseraeu and J. Arras, "Essais d'application de la dimension structurale en phylosociologie", *Vegetatio* IX, 1959, Dr. W. Junk, The Hague.
1983   JCE

Contents

Changes in nutrient cycling; The climax concept;
Challenges to the climax concept; Is succession
an orderly process?; Succession viewed as the
replacement of opportunist with equilibrium
species; The importance of succession for human
food production

List of Tables

CHAPTER I

Ecosystems

INTRODUCTION

The study of plants and animals in relation to their environments is known as *ecology*. The word ecology is derived from the Greek *"oikos"* meaning "house" or "place to live". So, literally, ecology is the study of organisms "at home" or "in their habitats". Although the word ecology has become part of the general vocabulary only in the last two decades, the science of ecology has been a recognised distinct field of biology for at least 70 years.

Its development followed attempts by naturalists in the eighteenth and nineteenth centuries to understand the distribution patterns of plants and animals. These early workers realised that the plants and animals of the world could be divided into major groups, otherwise known as *biotic associations* or *biomes*, according to the climatic regions they inhabited. For example, these included the tropical rain-forest and the hot desert biomes in which recognised associations of species occurred. However, this distributional approach could provide only a very limited amount of explanation. Consequently, efforts were made to replace it with a more functional concept to include plants and animals in their physical habitat working together as a system.

BASIC CHARACTERISTICS OF ECOSYSTEMS

1. Definition. A system of organisms functioning together with their non-living environment is known as an *ecosystem*. This concept is very broad and flexible so that it can be applied to any situation where organisms function together with their environment in such a way that there is interchange of materials between them, even if it is only for a very short time.

2. Scale. The largest and most nearly self-sufficient ecosystem we know about is the world as a whole which, of course, includes all the earth's living plants and animals interacting with the physical environment. This global ecosystem is often referred to as the "ecosphere" or "biosphere". The idea of the ecosystem can be

1

applied on a much smaller scale. For instance, we can think of forests, ponds, fields or even puddles as ecosystems.

3. Boundaries. In some cases habitats have clear boundaries, for example islands or woods, and therefore it is easy to delimit them and to study them as ecosystems. However, the boundaries of an ecosystem are often placed more arbitrarily, as in the case of a part of a forest or an area of the sea.

4. Basic features. All ecosystems have certain basic features of structure and function in common, no matter on what scale the concept is applied or whether the boundaries can be clearly defined or not. Most importantly, ecosystems all have (*a*) *biotic* (living), and (*b*) *abiotic* (non-living) components between which there is an exchange of energy and materials. Before these features are examined in detail it would be useful to consider some of the characteristics of systems in general.

CHARACTERISTICS OF GENERAL SYSTEMS

5. General systems theory. The idea of systems is used in many branches of science to help to understand or to explain inter-relationships. This approach, based on general systems theory, was devised in the early part of this century by L. von Bertanlanffy and has had a great influence on the development of thinking in academic subjects.

(*a*) *Definition.* A system can be defined as a set of objects or attributes (that is characteristics of an object such as its size or shape) linked by some relationship.

(*b*) *Examples of systems.* We are all familiar with the idea of systems in everyday life, for example the hot water system of a house, the electricity grid system and the railway system. Similarly, in biology we are used to talking about many types of system, including circulatory systems, excretory systems and reproductive systems.

(*c*) *The use of systems study.* The study of systems enables attention to be focused on working relationships between objects rather than on the individual objects themselves. Thus, in ecosystems we can study organisms functioning with their environment rather than just studying the individual types of plants and animals. The study of ecosystems not only provides a lot of information about the distribution and function of organisms, but also forms the basis of the management and conservation of the environment.

6. Closed and open systems. There are two basic types of system. These are, firstly, closed systems in which no energy or materials cross the external boundaries of the system (see Fig. 1(*a*)), and, secondly, open systems in which energy and materials do cross the boundaries of the system (*see* Fig. 1(*b*)).

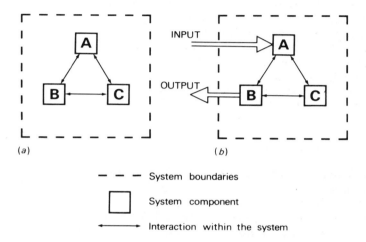

FIG. 1   *Two types of system. (a) Closed; (b) open.*

(*a*) The gains of energy and material to an open system are called *inputs*.

(*b*) The losses of energy and materials from an open system are called *outputs*.

(*c*) The exchange of energy and materials between the components within the system is known as *throughput*.

So, in the case of the water system of a house, water entering the house from the mains supply would be "input" whilst water leaving to the drains would be "output", and the water circulating in the pipes of the house would be "throughput".

With the exception of the universe, all natural systems, including ecosystems, are open systems. However, the degree of self-containment varies a great deal. For instance, a stream ecosystem could be thought of as being more open than a pond ecosystem, since in the former the water would be constantly flowing through, carrying materials away with it.

7. Steady-state. A very important characteristic of natural open systems is the fact that they tend towards a condition of balance or steady-state so that the systems components are all in harmony with each other. This balance is achieved by a process of self-regulation by which the components attempt to adjust themselves to any changes of inputs and throughputs of energy and materials.

For example, the number of animals living in a particular habitat will depend on the food supply in that ecosystem so that these two system components (numbers of animals and food supply) will be in balance. If for some reason, such as a decrease in rainfall, the amount of food for the animals diminishes, then fewer could be supported and these components would no longer be balanced. The population size of the animals would have to adjust to the new limited food supply either by starvation or migration. The number of animals present would be reduced until the population remaining could be supported by the food supply. In this way, a new balance would be achieved between these two system components.

8. Feedback. Feedback can take place in all types of system. It occurs when a change in one of the system components initiates a series of changes in the other components which eventually "feeds back" to affect the first component again. There are two types of feedback.

(a) *Negative feedback.* This is the most common and is a fundamental mechanism involved in achieving and maintaining balance or steady-state in ecosystems. Negative feedback has the effect of decreasing the rate of change in the component which originated the series of changes. Negative feedback leads to balance or steady-state.

An example of negative feedback could occur in a grassland ecosystem if the population of grazing animals is increased by animals emigrating from an adjacent area. In this way the system component of "grazing animals" is changed. However, if the number of animals increases greatly, the grassland will probably become overgrazed and eroded. Thus, there is a change in another component of the system, that is, the grassland component. The decreased food supply would have the ultimate effect of limiting the number of grazing animals in the ecosystem, so reducing the rate of increase in this system component and producing negative feedback.

(b) *Positive feedback*. This occurs far more rarely than negative feedback. In positive feedback a change in a system component causes a series of changes in the system which eventually lead to an acceleration in the rate of the original change. Positive feedback accelerates change and therefore tends away from balance or steady-state.

There are few examples of positive feedback in natural systems. However, it could occur if, for instance, a lake became polluted. The pollution could kill some of the fish resulting in a decrease in the "fish population" component of the ecosystem. The decaying bodies of the fish would contribute to the effects of the pollution and could cause the death of more fish. Thus, the rate of fish death would be accelerated producing positive feedback. When this occurs in natural systems it tends to feature as short bursts of destructive activity. Over the long term negative feedback and self-regulation tend to prevail.

COMPONENTS OF THE ECOSYSTEM

There are two basic components of the ecosystem: first, the abiotic part, which is non-living, and second, the biotic part which is living. Both of these components are equally important to the ecosystem because without one of them the system would not function.

9. The abiotic component. This includes all the factors of the non-living environment such as light, rainfall, nutrients and soil. These environmental factors not only provide essential energy and materials but they also play an important role in determining which plants and animals can inhabit an area. The detailed influences which the individual environmental factors exert on living organisms will be considered separately in V and VI.

10. The biotic component. This can be divided into three parts on the basis of function.

    (a) Producers.
    (b) Consumers.
    (c) Decomposers.

These three groups need to be considered in more detail.

11. Producers. These are the green plants which are capable of producing their own food. Because of this they are known as *autotrophs* (*auto* = self, *troph* = feeding). The green plants have the ability to use the energy of sunlight to manufacture carbohydrates from simple inorganic elements such as carbon, hydrogen and oxygen. This process is known as photosynthesis and can be summarised in the following way:

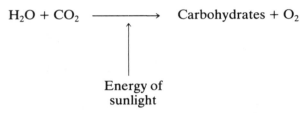

$$H_2O + CO_2 \longrightarrow Carbohydrates + O_2$$

Energy of
sunlight

Carbohydrates provide the basic building blocks from which other foodstuffs, such as proteins and fats, can be made if combined with other essential nutrients such as nitrates, phosphorus and potassium.

12. Consumers. These are animals which obtain their food by eating plants or other animals. Because of this they are known as *heterotrophs* (*hetero* = other, *troph* = feeding) There are four main types of consumer:

    (*a*) *herbivores* which eat only plants;
    (*b*) *carnivores* which eat only other animals;
    (*c*) *omnivores* which eat both plant and animal material; and
    (*d*) *detritivores* which eat only dead plant and animal material.

The consumers convert the materials of plant bodies to the materials of animal bodies. All the food energy obtained and used by the consumers comes from that originally made by the plants.

13. Decomposers. These are the organisms that promote decay. Decomposition is usually caused by micro-organisms such as the bacteria but it can also be accomplished by fungi. The decomposers break down the complex organic molecules which have been manufactured by the plants and animals. In this way the simple inorganic elements are liberated and can be reused by the plants.

    Decomposers are an essential part of ecosystems because without them the basic materials of life would become locked in complex molecules and further growth would be stopped.

Theoretically it is possible to have ecosystems which contain only producers and decomposers. However, the majority of ecosystems do have some consumers as well.

THE POND AS AN EXAMPLE OF A SIMPLE ECOSYSTEM

The basic components of ecosystems can be recognised in very different types of habitat and it would be useful to consider one as an illustration. The pond (*see* Fig. 2) is a good example of a small ecosystem because it has a recognisable unity and demonstrates the interrelationships between the biotic and abiotic components of the system very well. The pond is not only a habitat for plants and animals but the organisms themselves make the pond what it is. In a small pond the basic ecosystem components would be comprised of the following.

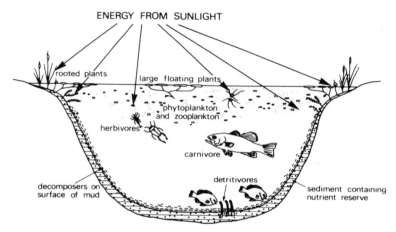

FIG. 2   *The pond as a simple ecosystem.*

14. Abiotic substances. These are the basic inorganic and organic compounds such as water, carbon dioxide and oxygen, together with the less abundant nutrients such as calcium and mineral salts. A small proportion of these vital nutrients would be in solution in the pond water and would be immediately available for use by the biotic component of the system. However, a much larger proportion would be present in sediments at the bottom of the pond. These nutrients in the mud form a reserve supply for the ecosystem.

15. Producers. In a pond there are usually two main types of producer organism:

(*a*) *large rooted or floating plants*, for example pond weeds and water lilies. These are generally confined to shallow water; and

(*b*) *minute floating plants*. These are usually algae and are called *phytoplankton* (*phyto* = plant, *plankton* = floating). These are distributed throughout the pond water where there is sufficient light energy for photosynthesis. These minute plants are not visible to the casual observer but if present in abundance the phytoplankton give the water a greenish colour. In most ponds the phytoplankton are much more important than the large plants are in the production of basic food.

16. Consumers. Many types of animal may be present in a pond ecosystem. For example, the herbivores may range from minute *zooplankton* (animal plankton) which feed on phytoplankton, to large herbivorous fish which feed on the pond weeds. Similarly, the carnivores can include a wide variety of animals such as predaceous insects and game fish. The carnivores may feed on the herbivores or on each other. Detritivores, for example aquatic worms, usually inhabit the surface of the mud at the bottom of the pond. These organisms subsist on organic detritus falling down through the pond water.

17. Decomposers. The aquatic bacteria and fungi are distributed throughout the pond but are especially prolific at the mud–water interface along the bottom of the pond where the dead bodies of plants and animals accumulate. In ponds dead organisms are generally broken up into pieces quickly, partly because of the action of decomposers. When the dead bodies are decomposed the nutrients in them are released for reuse by the plants.

TROPHIC STRUCTURE OF ECOSYSTEMS

18. Food chains. The transfer of food energy from the source in the plants through a series of organisms with repeated stages of eating and being eaten is known as the food chain. In the example of the pond ecosystem we have seen that there is a definite arrangement of the main biotic components to form a sequence of levels of eating. Food chains can be of a simple linear form as in:

plants $\longrightarrow$ herbivores $\longrightarrow$ carnivores $\longrightarrow$ decomposers

For example, in a marine ecosystem a food chain could be:

phytoplankton $\longrightarrow$ zooplankton $\longrightarrow$ whale $\longrightarrow$ bacteria

and in a grassland ecosystem a food chain could be:

grass $\longrightarrow$ vole $\longrightarrow$ stoat $\longrightarrow$ bacteria

Often there are more than four steps in a food chain. For instance, an ecosystem could contain a sequence of three types of carnivores feeding on each other.

plants $\longrightarrow$ herbivores $\longrightarrow$ carnivores $\longrightarrow$ carnivores
           (1)   (2)

$\longrightarrow$ carnivores $\longrightarrow$ decomposers
 (3)

However, it is rare to find more than six steps in a food chain. It is convenient to divide food chains into two basic sorts (a) grazing; and (b) detrital food chains.

(a) *Grazing food chain*. In this the plants are eaten live by the herbivores. This involves a fairly rapid and direct transfer of food energy from living plants to grazing animals and carnivores.

(b) *Detrital food chain*. In this dead plant material (detritus), such as dead leaves, is eaten by the detritivores. In terrestrial ecosystems these animals include the soil mites, millipedes and earthworms, and in aquatic ecosystems they include various worms and molluscs. The detrital food chain transmits energy to the other ecosystem components more slowly than the grazing food chain does. Dead plant material may remain in the system for a long time before it is consumed, although the action of the detritivores is often aided by the decomposers. Detrital food chains are generally more complicated than the grazing food chains.

These two types of food chain may exist separately or they may operate in conjunction with each other. The grazing and detrital food chains vary in importance in different types of ecosystem. For example, in forests the detrital food chain is often more important whereas in marine ecosystems the grazing food chain is usually more important. This contrast is not inherent in terrestrial and aquatic systems, as, for instance, in salt marshes the detrital food chains predominate but in grasslands the grazing food chains are more important.

19. Food webs. Although simple linear food chains can be found in many types of ecosystem, feeding relationships are frequently more complicated than this because the majority of animals consume a wide variety of food. Most herbivores eat many types of plant, while most carnivores eat several types of herbivore and other carnivores. Consequently the linear food chains interconnect to form food webs. An example of a food web for the grasslands of East Africa is shown in Fig. 3 (this diagram does not attempt to include all the animals involved in the web but illustrates just the main groups).

The grazing and detrital food chains often link up in food webs at the carnivore level. The patterns of feeding relationships in

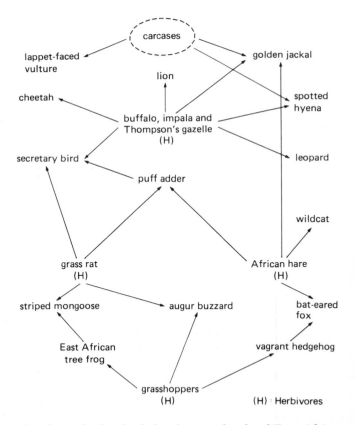

FIG. 3   *Part of a foodweb for the grasslands of East Africa.*

food chains and webs can be determined by several techniques such as by analysing the gut contents of animals and by introducing radioactive tracers to plants and monitoring their progress through the system.

20. Trophic levels. Organisms feeding at the same number of steps on a food chain or web from the autotrophs are said to be at the same trophic (feeding) level. Thus green plants occupy the first trophic level, the plant eaters occupy the second trophic level, the carnivores which eat herbivores occupy the third trophic level and so on. The consumers may be described also as primary, secondary, tertiary, etc., depending on which trophic level they are at. For example, in a food chain with five trophic levels there would be primary, secondary and tertiary consumers:

$$\text{plants} \longrightarrow \underset{\text{(primary)}}{\text{herbivores}} \longrightarrow \underset{\text{(secondary)}}{\text{carnivores}} \longrightarrow \underset{\text{(tertiary)}}{\text{carnivores}}$$

$$\longrightarrow \text{decomposers}$$

It is important to note that the idea of the trophic level is one of function not of population. One species can occupy more than one trophic level. For instance, in the food web shown in Fig. 3 the striped mongoose occupies the third trophic level when it eats grass rats but it occupies the fourth trophic level when it eats East African tree frogs. The arrows on Fig. 3 indicate the direction of energy flow.

BASIC LAWS OF ENERGY FLOW

Energy from the sun is the ultimate driving force of all ecosystems. The green plants do work with the energy of sunlight to collect nutrients from the soil and gases from the air to produce food. The food energy is passed through the system in the food chains and webs from one trophic level to the next. In this way energy flows through the system. Ecologists have traditionally looked at this energy flow in ecosystems in the same way as other scientists have examined energy flow in other physical systems. They have recognised that the energy may be present in several different forms.

21. Forms of energy. The energy available for use in ecosystems is present in several different forms or states. Four of these are most important.

(a) *Radiant energy*. This is the energy of light and is composed of a broad spectrum of electromagnetic waves radiating from the sun.

(b) *Chemical energy*. This is energy stored in chemical compounds. During photosynthesis, light (radiant energy) is used to manufacture complex carbohydrates. These may, in turn, be built up into other materials such as proteins and fats in the plant. During their progress through the trophic levels of the ecosystem the plant materials are converted to other complex substances to build the bodies of animals. This conversion uses up energy. When all these substances are broken down again, as in respiration, the energy is released. The complex compounds can, therefore, be thought of as an energy store.

(c) *Heat energy*. Heat energy results from the conversion of non-random to random molecular movements. This sort of energy is released whenever work is done. All types of work are included here, not only muscular contractions but also the complicated growth of organisms.

(d) *Kinetic energy (energy of motion)*. This is the energy which an organism possesses from its movement. The potential energy of chemical substances is converted to kinetic energy by means of movement when it is released to do work.

22. The measurement of energy. All forms of energy can be converted to heat equivalents. Consequently, the basic units of energy measurement most frequently used in ecology are:

(a) *The kilogram calorie (kcal)* which is the amount of heat needed to raise the temperature of 1 litre (1 kg) of water 1° (C) celsius (centigrade).

(b) *The gram calorie (cal)* which is the amount of heat needed to raise the temperature of 1 g of water 1° celsius (centigrade).

23. The laws of thermodynamics. The energy flow through an ecosystem operates within the framework of fundamental physical laws, namely the laws of thermodynamics. The first two of these laws are particularly important.

(a) *The first law* of thermodynamics states that energy cannot be created or destroyed. This means that it can only be transformed from one state to another. For example, light energy is transformed into chemical energy during photosynthesis.

(b) *The second law* of thermodynamics states that no transformation of energy from one state to another is ever 100 per cent efficient. There is always some loss as heat energy. This means, for instance, that when herbivores eat plants to get food for the growth and maintenance of their bodies, they will not be able to use all the food energy in the plant material. During the conversion of plant materials to animal materials there will always be some wastage of energy as heat.

24. The fate of energy entering the ecosystem. The first and second laws of thermodynamics dictate that all the energy of sunlight fixed as food by the plants must do one of three things:

(*a*) it can be passed through the ecosystem via the food chains and webs;

(*b*) it can be stored in the system as chemical energy in plant or animal materials; or

(*c*) it can escape from the system as heat or as outputs of material.

These three aspects need to be examined in more detail.

ENERGY FLOW AND THE STANDING CROP

25. Storage of energy in the system. This is the amount of plant and animal material present. The actual amount of living material contained in the ecosystem is known as the standing crop. It can be expressed in several ways but it is described most frequently as *biomass* (living material) per unit area measured as dry weight, ash weight or calorific value. Ecologists usually study the standing crop at each trophic level as this gives an indication of the pattern of energy flow through the system.

26. How the standing crop is determined by the pattern of energy flow. In most cases the amount of standing crop in each trophic level decreases with each step in the food chain away from the plants. This is for two reasons.

(*a*) *There are energy losses between trophic levels.* As we have noted already, the conversion of energy from one form to another is never 100 per cent efficient. Whenever the energy flows from one trophic level to the next, and the materials of one organism have to be converted to build the materials of another organism, some energy is lost. The transfer of energy between trophic levels involves large losses of energy.

(*b*) *There are energy losses within trophic levels.* All organisms must respire to live. Respiration involves the oxidation of carbohydrates to release energy for use in the body of the plant or animal. This process can be summarised by:

$$\text{carbohydrates} + O_2 \longrightarrow CO_2 + H_2O$$

$$\downarrow$$

ENERGY

In this way, energy is used up within each trophic level.

The combined action of (a) and (b) means that the flow of energy will decrease with each successive trophic level. There will be less energy available for use by the later steps in the food chain and so less biomass can be supported.

TROPHIC PYRAMIDS

Most ecosystems contain large numbers of food chains and complicated food webs. It is difficult to examine the patterns of energy flow in these individually, so generalisations are made to aid analysis. Organisms are usually grouped by their trophic levels and the relationships between the standing crops of each of these is considered. The amounts of living material at each trophic level can be shown diagrammatically as trophic pyramids (see Fig. 4). In these each bar represents a trophic level. The size of the bar is proportional to the amount of living material at that level. There are three basic types of trophic pyramid.

27. Pyramid of numbers. In this the numbers of organisms at each trophic level are noted. The normal pyramid, first studied by Elton in 1927, is one in which many small producers support a relatively large number of herbivores and carnivores (see Fig. 4 (a)). On occasions the number of producers is small compared to that of the consumers. For example, one or two trees could support a vast number of insects. Pyramids of numbers suffer from the major disadvantage that they make no allowance for the size of the organism. Producers ranging in size from minute plankton to the giant sequoia (redwood) trees of California are all dealt with in terms of the numbers present.

28. Pyramid of biomass. In this the weight of organisms present in

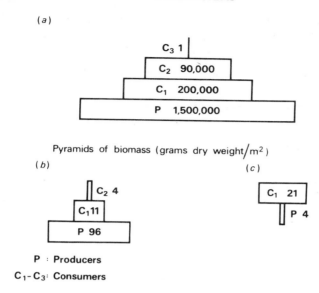

(a)

C₃ 1
C₂ 90,000
C₁ 200,000
P 1,500,000

Pyramids of biomass (grams dry weight/m²)

(b)

C₂ 4
C₁ 11
P 96

(c)

C₁ 21
P 4

P : Producers
C₁-C₃: Consumers

FIG. 4 *Examples of trophic pyramids. (a) Pyramid of numbers; grasslands in summer; (b) Wisconsin Lake; (c) English Channel.*

each trophic level is noted. Typically the weight of the producers exceeds that of the herbivores, which in turn exceeds that of the carnivores (*see* Fig. 4(*b*)). However, sometimes the opposite situation, known as an inverted pyramid, arises (*see* Fig. 4(*c*)). This pattern is found in two main situations: firstly, when measuring is done in a very restricted area, and secondly when the producer organisms are very small and have very rapid growth rates.

In general the smaller the organism, the greater its rate of metabolism (body chemistry) per gram of body weight. This trend is often known as the *inverse size-metabolic rate law*. It means that 1 g of small algae may be equal in metabolism to many grams of forest tree leaves. If the producers in the ecosystem are composed of very small organisms and the consumers are very large, the standing crop biomass of consumers may be greater than that of the producers at any one time. This situation can, of course, persist only if the plants can produce biomass at a very rapid rate because their carbohydrate production must be greater on average than that needed by the consumers.

Pyramids of biomass are useful measures of the energy stored at each trophic level since the differences in the calorific value of the materials in the ecosystem are not usually large. For instance, the energy value of most plant substances is 4 kcal per ash-free dry gram and that of most animal substances is 5 kcal per ash-free dry gram. The only materials which differ significantly from these calorific values are food stores, such as those present in seeds, which may have energy values of between 6 and 7 kcal per ash-free dry gram.

29. Pyramid of energy. In this type of pyramid the amounts of energy utilised by the organisms at each trophic level are evaluated for set sample areas (e.g. 1 m^2) over given periods of time (e.g. 1 year). This is a more sophisticated and potentially more accurate method than either the pyramid of numbers or the pyramid of biomass. However, it is difficult to work out and there are few examples of it.

TYPICAL PATTERN OF ENERGY FLOW THROUGH ECOSYSTEMS

Although the study of trophic pyramids is useful in providing an indication of the pattern of energy storage in the ecosystem we need to go further than this and examine the patterns of energy transfer between trophic levels as well as the losses of energy from the system. In order to do this we need to have a method of representing the flow of energy between the system components and across the system boundaries. For convenience, ecologists do this by grouping organisms into their trophic levels.

The amount of biomass at each trophic level can be shown diagrammatically as boxes. The flows of energy through the ecosystem can then be likened to the flow of water through pipes connecting the different components of the system (*see* Fig. 5). The pipes representing energy flow are of various widths proportional to the amounts of energy involved. This idea, known as the *hydraulic analogy*, was first developed in 1956 by an American ecologist, Howard Odum. It enables us to demonstrate the pattern of energy flow in an ecosystem.

30. The main pathways of energy flow. These are as follows.

(*a*) Energy enters the ecosystem as sunlight but not all of this energy is used in photosynthesis. Only about half of the average sunlight impinging upon green plants is absorbed by the photosynthetic mechanism and only a small proportion of the absorbed energy (about 1 to 5 per cent) is converted into food (chemical energy). The rest of the energy escapes from the system as heat. This is shown draining away from the system in Fig. 5. Some of the energy made into food by the plants is used in their respiration. This process gives off heat energy which is lost from the system.

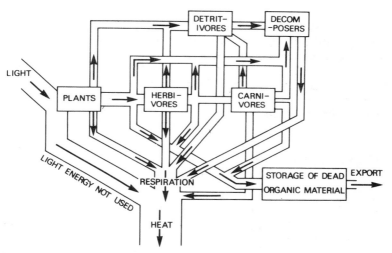

FIG. 5 *The hydraulic analogy of energy flow through an ecosystem.*

(*b*) Energy stored as plant material may pass through the feeding chains and webs via the herbivores and detritivores. As we have noted already large losses of energy occur between trophic levels; therefore the flow of energy decreases towards the later stages of the food chains. Typically, the herbivores store about 10 per cent of the energy provided by the plants. Similarly the carnivores store only about 10 per cent of the energy provided by their prey.

(*c*) If the plant material is not consumed it may be stored in the system, pass to the decomposers or be exported from the system as dead organic matter.

(*d*) The organisms at each of the consumer levels and also at the decomposer level use some energy for their own respiration and so liberate heat which drains away from the ecosystem.

(*e*) Because the ecosystem is an open system some organic material may be imported across the system boundaries. This is not shown on Fig. 5 but could take place, for example, by animals moving into the area or by water weeds being carried downstream to a lake.

31. Disadvantages of the hydraulic analogy. The hydraulic analogy suffers from two main disadvantages which both derive from its simplicity:

(*a*) In reality it is difficult to group organisms into trophic levels. One species may have several roles in an ecosystem and may operate at several different trophic levels.

(*b*) In nature energy flow paths are rarely simple. Energy flows through the system by a complex of encounters between species leading to interaction between trophic levels. Frequently there are loops of flow rather than straightforward direct pathways. For instance, one species may feed on the faeces of another. In this case the energy in the faecal material does not go to the decomposers but gets taken back into the food chains at a lower trophic level.

Although the hydraulic analogy gives a neat impression of the functioning of an ecosystem it should be remembered that it is an extreme abstraction of reality and as such its practical use is limited.

PROGRESS TEST 1

1. Define the "ecosystem". **(1)**
2. Name the two basic components of all ecosystems. **(4)**
3. What are the differences between closed and open systems? **(6)**
4. Describe negative feedback and positive feedback. **(8)**
5. Name the four types of heterotroph. **(12)**
6. What are phytoplankton? **(15)**
7. What is the main difference between grazing food chains and detrital food chains? **(18)**
8. Define the term trophic level. **(20)**
9. What are the main forms of energy? **(21)**
10. Why are the first and second laws of thermodynamics relevant to ecology? **(23)**

11. Describe briefly the three types of trophic pyramid. (27–29)

12. How may energy be lost from ecosystems? (30)

13. What are the disadvantages of the hydraulic analogy? (31)

CHAPTER II

Nutrient Cycling

BASIC IDEA OF NUTRIENT CYCLING

The flow of energy in ecosystems is one-way. The energy of sunlight fixed as food by plants is dissipated quickly through the food chains and ultimately escapes from the system. In contrast, the nutrients which are needed to produce organic matter are circulated round the ecosystem and reused several times. When the bodies of plants and animals are decomposed by the action of bacteria and fungi the nutrients in them are liberated to the abiotic environment and form a nutrient pool or reservoir.

For example, in terrestrial ecosystems the nutrients are usually liberated into the soil. The nutrients can be taken up from the reservoir to be reused by the plants. In this way the nutrients re-enter the ecosystem and can be passed through it again. In Fig. 6, a nutrient cycle is shown superimposed on an energy flow diagram to demonstrate how the one-way flow of energy drives the nutrient cycle.

It is through the action of the nutrient cycles that the organic and inorganic components of the ecosystem are inextricably linked together and are so mutually interdependent that it is often difficult to separate them in nature. The two fundamental processes of energy flow and of nutrient circulation are characteristic of, and common to, the whole biosphere. In both processes the green plant provides the vital link between the abiotic and biotic components of the ecosystem. It is the point at which both the energy and the nutrients enter the biotic part of the system.

NUTRIENTS REQUIRED

1. Types of nutrients. Of the 92 elements known to occur in nature some 30 to 40 have been identified as being essential for living organisms. The nutrients, also known as biogenic salts, can be divided into two main groups.

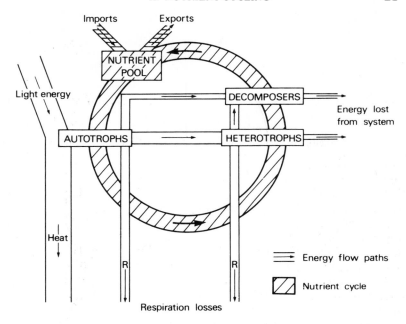

FIG. 6   *A biogeochemical cycle (shaded circle) superimposed on a simplified energy-flow diagram.*

(*a*) *Macronutrients.* These are required in relatively large quantities and play a key role in the formation of protoplasm (living material). The most important nutrients in this group are hydrogen, carbon, oxygen and nitrogen, which together account for over 95 per cent of the dry weight of living material. These four nutrients are obtained directly or indirectly in gaseous form from the atmosphere. The macronutrients include also other nutrients which are needed in smaller amounts such as potassium, sulphur and phosphorus.

(*b*) *Micronutrients or trace nutrients.* These elements are required in very minute amounts but they are still essential for life. At least ten micronutrients are required by plants. Many of these micronutrients, which include iron, copper, zinc and boron, have their initial sources in rocks from which they are released by processes of weathering.

2. The relative abundance of nutrients in organisms.

(*a*) *Amounts present in biomass.* Although all the natural elements are capable of being absorbed by plants only three, carbon, oxygen and hydrogen, are usually found in very large quantities in living material. Of these, hydrogen and oxygen together as water will account for much of the total weight of all life forms. For instance, the proportion of water in wood is over 50 per cent, that in most vertebrates over 66 per cent, while that in certain marine invertebrates is over 99 per cent. In consequence, most other elements, including some of the macronutrients, are found in only small amounts in biomass.

(*b*) *Amounts taken in by organisms.* Most elements enter living organisms in either a gaseous state (e.g. O_2, CO_2) or as water-soluble salts (e.g. NaCl). The majority of plants are fairly indiscriminate in their nutrition, taking in elements and compounds from the environment in more or less the proportions in which they exist in the surroundings. However, for a variety of complicated environmental and physiological reasons there will be some inevitable differences between the intake of specific elements in different types of organism. These differences arise from factors such as the chemical characteristics of the nutrient involved and the metabolism of the individual plant or animal species.

(*c*) *Selective chemical enrichment.* In some cases organisms can accumulate elements in their bodies to concentrations in excess of those prevailing in their surroundings. This phenomenon is known as *selective chemical enrichment*. It features most frequently in plants but occurs also in simple aquatic organisms. For example, many marine organisms are capable of extracting large amounts of calcium carbonate and silica from sea water, while oysters can accumulate copper in their bodies to concentrations in excess of 200 times that in the adjacent water.

3. The use of the macronutrients. The three main macronutrients (carbon, hydrogen and oxygen) act as major components in fats and carbohydrates and also help to form the basic cell structure of plants and animals. In plants the cell walls are composed mainly of a rigid substance, cellulose, which is built from these three nutrients in a weight ratio of carbon 7.2 : hydrogen 1 : oxygen 8. Carbon, hydrogen and oxygen together with nitrogen (the fourth most important macronutrient) form the basis of proteins.

The addition of phosphorus to these four provides the building blocks of many of the nucleic acids which form the genetic blueprint of cells. This combination of nutrients forms the basic material of much of the cell structure. Phosphorus is essential also for the transformation of energy through the cells.

Other macronutrients are required in smaller quantities. For example calcium is used for strengthening the cell walls, and sulphur is needed for the formation of amino acids which are used to build proteins.

4. The use of the micronutrients. Micronutrients are often essential components of structures in the cell. For example, magnesium is needed in minute amounts to produce chlorophyll, the green pigment of plants, which is used to absorb light energy during photosynthesis.

Frequently, the micronutrients are required to manufacture enzymes. These are proteins which act as organic catalysts to speed up chemical reactions in the cell. Many enzymes need specific chemical elements in their structure. For example, the enzyme nitrate reductose, which helps to change nitrate into ammonia, needs trace amounts of molybdenum in its molecules.

5. The problem of nutrient use in animals. Animals not only have more varied nutritional requirements than plants but also their metabolism involves different chemical reactions. In general, animal cells have a more limited capacity to synthesise organic compounds. Indeed their metabolism often cannot be maintained unless they are provided with a wide selection of readily available organic foods such as proteins (or amino acids), vitamins and fats.

These foods are obtained from those already synthesised by plants. Heterotrophs acquire them either directly by eating the plants themselves, or indirectly by eating other animals. Therefore, although identical elements circulate in both plants and animals, the nutrients pass from the first trophic level to the second as readymade organic compounds rather than as simple substances.

6. Non-essential elements. Many elements that have no known biological function circulate between the biotic and abiotic components of ecosystems. For example, large quantities of minerals, such as silicon, pass back and forth to the abiotic environment. These non-essential elements circulate as a result of the indiscriminate absorption of elements by plants.

This characteristic may be of great ecological importance if the non-essential elements occur in quantities that are chemically toxic, or if they react chemically in the soil in ways which render the essential elements unavailable to the plant.

CHARACTERISTICS OF BIOGEOCHEMICAL CYCLES

7. Introduction. It is clear that the functioning of all ecosystems depends on the circulation of nutrients. If the nutrients are not circulated, available supplies become exhausted and growth would be limited. Despite this importance, little work was done to determine the pathways of nutrient cycling until the 1930s. However, the main routes and mechanisms of most cycles have been deciphered now.

In the majority of cases it is useful to consider the circulation of nutrients on a global scale since there will always be large inputs and outputs of nutrients from ecosystems at a local level. For example, nutrients are constantly leaking from all terrestrial ecosystems in the drainage water. In return, new supplies enter the system in rainwater and by the weathering of rocks.

8. Basic features of biogeochemical cycles. The global scale nutrient cycles are known as *biogeochemical cycles* ("bio" for the living part of the system, "geo" for the non-living environment). All biogeochemical cycles involve interaction between soil and atmosphere.

The exchanges of nutrients in the cycles require the presence of a wide variety of living organisms whose patterns of birth, life and death all encourage the movement of the elements through the ecosystems. Because of this one could expect the cycles to take place at every level in the biosphere. Indeed, some processes in the biogeochemical cycles take place at high altitudes in the atmosphere while others have been detected in bedrock several thousand metres below the surface where micro-organisms live off oily liquids. However, most exchanges take place in the immediate contact zone between the lower atmosphere and the upper parts of the land and oceans where life is abundant and the annual turnover of materials is enormous.

Each biogeochemical cycle has two main parts.

(*a*) A reservoir pool which is a large slow moving non-biological component and is inaccessible to organisms.

(*b*) An exchange pool or nutrient pool which is a smaller more active portion where the nutrient is exchanged between the biotic and the abiotic parts of the system.

9. Types of biogeochemical cycle. The cycles can be divided into two groups on the basis of where their reservoir pools are.

(*a*) *Gaseous cycles.* In these the reservoir pool is the atmosphere. The nutrients enter and leave the biosphere in gaseous form.

(*b*) *Sedimentary cycles.* In these the reservoir pool is the earth's crust. Nutrients enter the biosphere from weathered rock and leave it as sediments.

10. The stability of biogeochemical cycles. The cycles exhibit contrasts in the completeness of the exchange of the element between abiotic and biotic parts of the ecosystem. Generally, gaseous cycles are more complete than sedimentary ones. In nature both types of nutrient cycle are assumed to be stable but the sedimentary ones are more susceptible to disruption by human interference.

(*a*) Most gaseous cycles, such as those for carbon, oxygen and hydrogen, self-adjust quickly to local variations because of the large atmospheric reservoir pool. For example, local increases in carbon dioxide concentrations produced by burning fuels tend to be dissipated rapidly by air movements and increased uptake by plants. In this way, negative feedback controls operate.

(*b*) Sedimentary cycles, such as those for sulphur and phosphorus, are less perfect and are more easily disrupted by local activities. Local variations in the reservoir pool of the earth's crust are less easily adjusted because this pool is relatively inactive and immobile compared with the atmosphere. Consequently, if some portion of the nutrients in the exchange pool are lost to the reservoir pool of the earth's crust they become inaccessible to organisms for a long time.

11. The rates at which the nutrients are cycled. These vary a great deal both spatially and temporally. The most important aspects influencing the rate of cycling are as follows:

(*a*) *The nature of the element involved.* Some nutrients are cycled at faster rates than others because of their chemical characteristics and the way they are used by organisms. Nutrients in gaseous cycles are usually cycled faster than those in sedimentary cycles.

(*b*) *The rate of growth of plants and animals*. These will affect the rate of uptake of the nutrient and the movement of it through the food webs.

(*c*) *The rate of decay of organic matter*. This is primarily a function of climate and soil type. In favourable soils, such as those in warm humid environments, decomposer organisms exist in densities of several million per gram. This enables biomass to be decayed quickly resulting in the rapid release of nutrients. In contrast, the bacterial and fungal populations of cold, wet soils will be low and decomposition will be slow. Organic matter may accumulate in these situations because of its slow destruction rate. The acidity of the soil is also important in influencing the rate of decay. In general, alkaline soils promote the rapid decomposition of organic matter whereas acid soils, which are unfavourable for bacteria, discourage rapid decay.

(*d*) *The activities of man*. Human activities have a great influence on the rate of nutrient cycling. For example, agriculture and deforestation affect the rate at which minerals are washed from the soil by rainwater and the combustion of fossil fuels liberates sulphur and carbon dioxide to the atmosphere.

THE CARBON CYCLE

This is an example of a gaseous biogeochemical cycle because the main reservoir of carbon is the carbon dioxide of the atmosphere. Although this reservoir comprises only about 0.03 per cent of the atmosphere by volume, carbon is the most important single element of organic chemistry and is cycled regularly between the biotic and abiotic parts of ecosystems.

The carbon cycle has many features in common with other biogeochemical cycles but it shows the greatest resemblance to the flow of energy through an ecosystem. In general, the carbon cycle is a complete, well-buffered system and has adequate feedback controls to ensure a plentiful supply of carbon for organic growth. The main pathways of the cycle are shown in Fig. 7.

12. Uptake of carbon dioxide by plants. Almost all carbon enters the biotic part of ecosystems by direct uptake as carbon dioxide by plants. The carbon dioxide is used to manufacture sugars and carbohydrates in photosynthesis.

Some of these substances are incorporated into plant structures

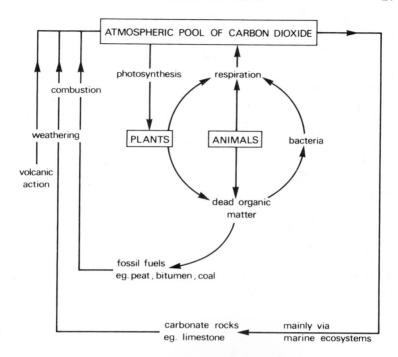

FIG. 7   *The carbon cycle.*

while others are used for respiration and the carbon dioxide in them released to the atmosphere. Approximately half of the assimilated carbon will eventually be added to the soil as decomposing organic matter.

The total quantities of carbon assimilated by plants are very large. It is estimated that 3 per cent of the total carbon dioxide content of the atmosphere is taken up annually.

13. Passage of carbon through food chains. The carbon assimilated in plant tissues may be passed through the food chains of ecosystems. Long food chains may expand the cycle considerably. Carbon will be returned to the atmosphere as carbon dioxide from respiration at each trophic level. Eventually the dead organic matter will either be stored in the system or pass to the decomposers.

14. Dead organic matter.

(a) Dead organic matter forms a subsidiary pool of carbon in the soil. Most of the carbon in the soil is derived from the decomposing surface litter of plants (leaves, twigs etc.) but important contributions come from decaying plant roots and dead animals.

(b) The rate and speed of breakdown of the organic matter and hence the rate of carbon release, varies a lot according to the nature of the organism. For example, trees decay less quickly than soft annual plants. In general carbohydrates, fats and proteins decompose most rapidly. Structural materials like cellulose take longer to decompose. Some materials are very resistant to decay. These form the organic component of the soil, known as humus. This continues to decompose very slowly.

(c) Both the organisms which cause decay, such as the bacteria, and the detritivores, such as the earthworms which help to break up the dead material, will release carbon dioxide back to the atmosphere by their respiration. On an annual basis the cycling of carbon through an ecosystem in this way is very complete. There will be some storage in the system but most carbon will be respired back to the atmosphere.

15. Fossil fuels.

(a) If conditions are suitable some of the dead organic matter may form organic deposits such as peat and coal. These deposits form an additional reservoir of organic carbon. The carbon in them will be out of circulation for a long period but can be returned to the atmosphere if the fuels are burnt.

(b) The increased use of fossil fuels in recent years has resulted in the rapid return of a lot of carbon dioxide to the atmosphere. Although some of this will be removed by natural processes, evidence suggests that the carbon dioxide concentration of the atmosphere is increasing and that an imbalanced situation is being created in the carbon cycle. Various consequences of this have been suggested. For instance, altering the radiation budget may cause a gradual warming of the world's climates.

16. Interchange between the atmosphere and the sea.

(a) Carbon dioxide can dissolve readily in sea water. Consequently, if the atmospheric concentration of carbon dioxide increases, much will be absorbed by the world's oceans. In spite of this buffer, about half of the carbon dioxide produced by fossil fuel combustion is known to stay in the atmosphere. This is because the oceans take about 200 years to be mixed sufficiently to absorb the surplus carbon dioxide.

(b) Carbon dioxide reacts with water to form carbonates, especially calcium carbonate ($CaCO_3$). This material can be used for shell construction in such animals as clams, oysters, some algae and some protozoa. After the death of the animal the calcium carbonate may dissolve or it may be incorporated into sediments which may eventually form carbonate rocks such as limestones and dolomites. Carbonate rocks act as an environmental buffer, storing the carbon for millions of years. It may be released again eventually as carbon dioxide, given off during the process of weathering.

17. Volcanic action. Carbon dioxide is emitted continuously from volcanic vents. This activity provides a fresh supply of carbon to the atmosphere from deep in the interior of the earth.

THE NITROGEN CYCLE

This is an example of a very complex gaseous cycle (*see* Fig. 8). The nitrogen cycle is probably the most complete of all the nutrient cycles since it has many self-regulating feedback mechanisms. Nitrogen is an element of special interest because it is a principal component both of living things and also of the atmosphere of which it forms about 80 per cent by volume. As in the case of the carbon cycle, the main reservoir pool is the atmosphere, and the exchange pool operates between organisms and the soil. However, unlike carbon, nitrogen cannot be taken into most plants directly from the air. It has to be made into chemical compounds, such as nitrate, before it is available to the exchange pool. This involves many complicated mechanisms.

18. The conversion of atmospheric nitrogen to nitrate. This can occur in two main ways.

(a) *By the action of specialised organisms.* These are mostly bacteria, algae and fungi, bacteria being the most important. They can operate either by themselves in the soil (e.g. *Azotobacter*) or in an association with a plant, especially those of the legume family such as clover. The association between a nitrogen-fixing bacterium (e.g. *Rhizobium*) and a legume forms

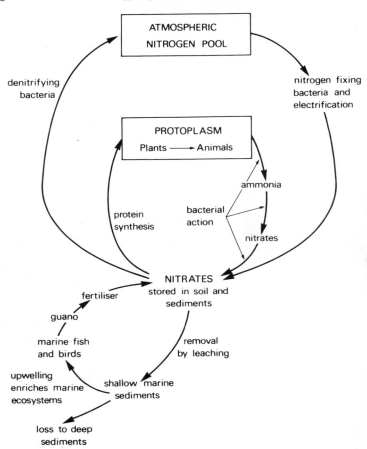

FIG. 8   *The nitrogen cycle.*

swellings of the roots, known as root nodules. Nitrate is formed in these and is either used by the plant or exudes to the soil. Because of this, legumes are very important in crop rotations to maintain soil fertility.

(*b*) *By the action of lightning.* Electrical activity during thunderstorms fixes atmospheric nitrogen as nitrate. This type of fixation is far less important than that due to biological action and accounts for only a small percentage of the total nitrate fixed for use in ecosystems.

19. The passage of nitrate through the food chains. Nitrate is used by plants for protein synthesis.

It is either absorbed directly from the root nodules, if the plant has them, or it is taken in from the soil. The nitrate assimilated to form plant proteins may pass through the ecosystem via the food chains.

Some loss of nitrogen occurs within each heterotrophic level as nitrogenous substances are excreted with the urine and faeces. Ultimately the nitrogen content of the organic matter is returned to the abiotic part of the ecosystem when plants, animals and waste products decompose.

20. The conversion of proteins to nitrate in the soil.

(*a*) Nitrogen in living bodies is present in large energy-rich molecules. The breakdown of proteins and other nitrogenous substances to nitrate provides energy for micro-organisms. This process of release of potential energy is called *deamination*. It involves several stages, each accomplished by specialised bacteria.

(*b*) Proteins are first broken down to amino acids from which mineral elements may be released. This process, known as *mineralisation*, involves the action of a wide range of bacteria and fungi.

(*c*) The amino acids are then decomposed further to produce ammonia (NH_4). Nitrifying bacteria, such as *Nitrosomonas*, use the potential energy of this compound and convert it to nitrite (NO_3).

(*d*) Nitrite may be toxic to plants if present in excess. However, soil nitrite is usually converted to nitrate (NO_4) rapidly by bacteria. The nitrate is chemically stable and may be reused by plants, thus completing a cycle within the exchange pool.

21. The conversion of nitrates to atmospheric nitrogen. If nitrates are not absorbed by plants they may be lost from the exchange pool by being broken down by *denitrifying* bacteria which release nitrogen to the atmosphere. Denitrifying bacteria live in places without oxygen such as estuaries, bottoms of fertile lakes, parts of the sea floor and waterlogged soil.

The denitrifying bacteria use energy to complete the transformation of nitrate to nitrite, thence to ammonia and eventually to free nitrogen.

22. The removal of soil nitrates by leaching.

(*a*) If nitrates are not absorbed by plants they may be washed downwards through the soil by heavy rains (a process known as *leaching*). In this way the nitrates may enter the river systems and ultimately be lost to shallow marine sediments.

(*b*) In certain areas, such as off the coast of Peru, upwelling currents enrich marine ecosystems. The nitrate may enter marine food chains and be returned to land in the droppings of seabirds. These droppings, known as guano, were once a major world supply of fertiliser.

(*c*) The use of artificial nitrate fertilisers has accelerated the loss by leaching of nitrate from agricultural areas greatly and has created a potential imbalance in the system.

(*d*) If the nitrate in shallow marine sediments is not recycled as guano it may be lost to deep sediments in which case it will be inaccessible to the exchange pool for millions of years. This loss will be compensated for slightly by the release of nitrogen gases from volcanoes.

THE PHOSPHORUS CYCLE

This is an example of a simple sedimentary cycle which is disrupted easily. The reservoir pool of phosphorus is in crystalline phosphate rocks and the exchange pool involves cycling between organisms, soil and shallow marine sediments (*see* Fig. 9). In common with the majority of biogeochemical cycles, the phosphorus cycle has no atmospheric phase. Phosphorus is relatively rare in nature but it is essential for plant and animal growth. As though in response to this, organisms have evolved ways of accumulating phosphorus in their living tissues at concentrations much greater than occur in the inorganic environment. The form in which phosphorus is found naturally in the environment is phosphate. This can be either as organic or inorganic compounds and may be present either in soluble or insoluble forms.

23. The passage of soil phosphate through food chains.

(*a*) The organic phase of the phosphorus cycle is very simple. Supplies of phosphate may be present in the soil either as organic compounds or as mineral salts. Both of these may be absorbed in solution by plants.

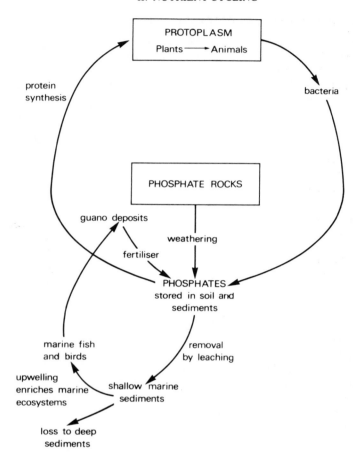

FIG. 9   *The phosphorus cycle.*

(*b*) The absorbed phosphate is used for protein synthesis by plants and may be passed through the food chains. When the plants and animals die the decomposition of their bodies and excretory products by bacteria and fungi releases phosphate back to the soil.

(*c*) Some inorganic phosphate may be immobilised temporarily in the bodies of micro-organisms which need it as food and therefore actively compete for it, often to the detriment of plants. The immobilised phosphate will become available again after the micro-organisms die.

The amount of phosphate available for life depends primarily on the rate at which it moves through the organic phase.

24. Leaching of phosphates from the soil.

(*a*) Most phosphate in the soil will be reabsorbed by plants quickly. Some will become bound tightly to clay minerals, such as kaolinite, and will be held in the soil. However, as in the case of most nutrients, some leaching will occur. Most of the phosphate carried from the land to the sea in the drainage water will eventually be absorbed into oceanic sediments and so lost to terrestrial ecosystems.

(*b*) One important route for the rapid return of phosphorus from shallow marine sediments is via the guano deposits. These are even more important in the case of phosphate than nitrate. However the return of phosphorus in this way tends to be very localised and it is estimated that it accounts for less than 3 per cent of that lost from land. If the phosphate is not recycled by upwelling currents it may be lost to deep sediments and incorporated into phosphate rocks.

25. The liberation of phosphate from rocks.

(*a*) The depletion of phosphorus from the exchange pool is compensated very slowly in nature by the release of the element from phosphate rocks. This occurs mainly by the processes of erosion and weathering, but is augmented by volcanic action. This return from the sediment phase is slow and unreliable which means that the cycle can be disrupted easily. The self-regulating devices characteristic of gaseous cycles are absent.

(*b*) Phosphate rocks are mined and used to manufacture phosphate fertilisers. The phosphorus in them is rapidly lost from the exchange pool by leaching. In this way man hastens the rate of loss of available phosphate and makes the cycle less perfect. This activity could lead to serious deficiencies in phosphorus supplies for agriculture in the future.

THE SULPHUR CYCLE

This is an example of another sedimentary cycle, but unlike the phosphorus cycle this one also has an atmospheric phase. The main reservoir pool of sulphur is held in sulphate rocks, such as gypsum, but in addition a small amount is present in the atmosphere. Not as much sulphur is required by ecosystems as carbon, nitrogen or phosphorus, nor is it a limiting factor in the growth of organisms as often as these other nutrients are. Nevertheless, the sulphur cycle (*see* Fig. 10) is a key one in the general pattern of growth and decomposition. Many of the steps

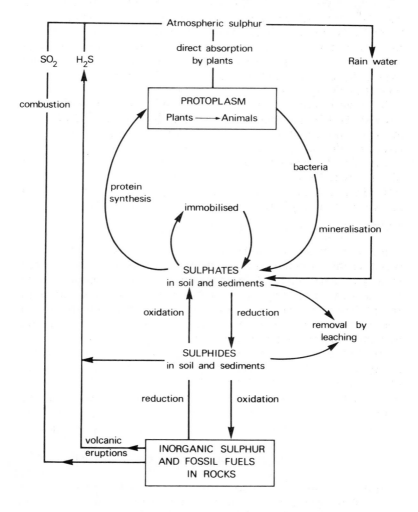

FIG. 10 *The sulphur cycle.*

in the cycle involve the activity of specialised micro-organisms, each conducting particular chemical reactions.

26. The passage of sulphate through the food chains. Sulphur is required by organisms for the synthesis of proteins and vitamins. Plants acquire most of their sulphur as sulphate (SO_4) from the soil. In addition, small amounts may be absorbed directly from the atmosphere as sulphur dioxide (SO_2).

The sulphate in plants may pass through the food chains or it may pass directly to the decomposers when the plant dies. In either case, the sulphate will be liberated back to the soil by the action of the bacteria and fungi which break down proteins.

Some of the sulphate in the soil may be immobilised temporarily by being incorporated into the bodies of micro-organisms which compete with the plants for available supplies. This activity prolongs the organic phase of the cycle.

27. Loss of sulphate from the soil. If the sulphate is not reabsorbed by the autotrophs it may be lost from the soil. This occurs in two main ways.

(*a*) Steady losses occur through leaching especially in soils poor in organic molecules.

(*b*) The sulphates may be converted to sulphides by reduction (the addition of hydrogen). Oxygen is released in the process. This is brought about by reducing bacteria, such as *Desulfovibrio desulfuricans*, which live in anaerobic conditions (places without oxygen) such as waterlogged soils and estuarine muds. If hydrogen sulphide (H_2S) is formed, this may escape to the atmosphere. Alternatively, if iron is present, the hydrogen sulphide may be converted to iron sulphide which is highly insoluble.

28. The sulphate reservoir in rocks.

(*a*) The sulphides in sediments and soils contain potential chemical energy. Chemosynthetic bacteria, which can release and use this energy, convert the sulphides to sulphates by oxidation (the addition of oxygen).

Some of this sulphate may be returned to supplies in the soil but some will be incorporated into sulphate rocks and inorganic sulphur in fossil fuels, thus entering the reservoir pool of the cycle.

(*b*) The most important natural route for the return of sulphate from the reservoir pool is by erosion and weathering. The inorganic sulphate from the rock is reduced to sulphides by the action of bacteria. The sulphides are in turn oxidised to form sulphates in the soil and so the sulphur becomes available to plants again.

(c) The sulphur in the reservoir pool may be released to the exchange pool as hydrogen sulphide given off to the atmosphere during volcanic eruptions. An increasingly important return route for sulphur from the reservoir pool is by the combustion of fossil fuels. When these are burnt the sulphur content is oxidised and enters the atmosphere as sulphur dioxide (SO_2).

29. The atmospheric phase.

(a) The atmosphere forms a significant additional reservoir for sulphur, which can enter the atmosphere as hydrogen sulphide, produced either by volcanic action or by the action of reducing bacteria, or as sulphur dioxide from the combustion of fossil fuels.

The release of sulphur dioxide by combustion has greatly increased the concentration of atmospheric sulphur especially in urban and industrial areas. This has created a potential imbalance in the cycle.

(b) Once in the atmosphere, the sulphur compounds are usually oxidised fairly rapidly to form sulphite (SO_3) and sulphate (SO_4). These processes are the result of chemical reactions in the air and do not involve organisms. Some sulphur compounds will be reabsorbed by the leaves of plants, some will be deposited dry on the surfaces of the earth and some will be washed out of the atmosphere by rainfall.

The sulphite and sulphate can combine with rain water to form sulphuric acid (H_2SO_4). In small quantities this is a useful way of returning sulphate to the soil. However, if present in excess, the sulphuric acid can cause detrimental effects by altering the conditions in the ecosystem.

PROGRESS TEST 2

1. What are micronutrients? (**1**)
2. Explain the term selective chemical enrichment. (**2**)
3. What are macronutrients used for in organisms? (**3**)
4. Name the two parts of biogeochemical cycles. (**8**)
5. Describe the main differences between gaseous and sedimentary biogeochemical cycles. (**10**)

  6. Which factors influence the rate of nutrient cycling? **(11)**

  7. What is the role of fossil fuels in the carbon cycle? **(15)**

  8. How may nitrate be formed from atmospheric nitrogen? **(18)**

  9. What is deamination? **(20)**

  10. Why is the phosphorus cycle unstable? **(24, 25)**

  11. What is the function of reducing bacteria in the sulphur cycle? **(27)**

  12. How can sulphur enter the atmospheric phase of its nutrient cycle? **(29)**

CHAPTER III

Productivity

THE CONCEPT OF PRODUCTIVITY

1. Biological productivity. The flow of energy and the circulation of nutrients are both essential for the growth of organisms and the functioning of ecosystems. Both of these processes are critical in determining the rate of production of new organic material, otherwise known as the *biological productivity*. It is important to note that productivity is the rate of production of biomass and must not be confused with the *standing crop* which is the amount of biomass present at one time in the ecosystem. The study of biological productivity is a vital part of ecology because it indicates the efficiency of different types of ecosystem and has far-reaching implications for improving production in the "man-made" ecosystems of agriculture.

2. Basic types of productivity. Productivity can be divided into two main types.

(*a*) *Primary productivity*, which refers to the production of new organic matter at the autotroph level (plants).

(*b*) *Secondary productivity*, which refers to the production of new organic matter at the heterotroph level (animals).

Each of these can be divided further into gross and net productivity.

(*c*) *Gross productivity* is the total amount of organic matter produced.

(*d*) *Net productivity* is the amount of organic matter left after some has been used to provide energy for respiration.

Therefore, we can refer to gross primary productivity, which is the total amount of carbohydrates produced in the plants by photosynthesis, and we can refer to net primary productivity, which is the amount of organic matter left after some has been used for respiration:

net primary productivity = gross primary productivity
− respiration losses.

Similarly, we can refer to gross secondary productivity, which is the total amount of organic matter assimilated by the animals from their food, and we can refer to net secondary productivity, which is the amount of organic matter left after some has been used for respiration:

net secondary productivity = gross secondary productivity
— respiration losses.

3. Units of measurement. Since productivity is the rate of production of organic material, its units of measurement must include a time-period. Different units have been used by various workers resulting in some confusion because their data are not always directly comparable. However, all the units employed include three basic measurements:

(a) a measure of biomass, e.g. dry weight, numbers of individuals, kcal;

(b) a measure of area, e.g. square metre (m^2), hectare;

(c) a measure of time, e.g. day, year.

The most frequently used combination is grams of dry weight per square metre per day ($g/m^2/day$). Additional care must be taken when comparing productivity measurements from different workers because production rates may have been averaged over contrasting time-scales. For example, the results of one worker expressed as $g/m^2/day$ may be production rates per day averaged over a whole year, whereas the results presented by another worker for the same ecosystem, again expressed as $g/m^2/day$, may be production rates per day averaged over the few months of the growing season. Obviously the figures will be very different.

BASIC PROCESSES INVOLVED IN PRIMARY PRODUCTIVITY

Very few detailed quantitative studies of primary productivity have been conducted, but those which have reveal very slow rates.

Net primary productivity is determined by the relative rates of photosynthesis (producing carbohydrates) and respiration (using up carbohydrates). In order to understand the factors which limit and control primary productivity, it is necessary to examine these two basic processes in more detail.

4. The process of photosynthesis.

(a) Only a small fraction of the light energy available to green plants is actually used in photosynthesis. A lot of light is reflected, passes through the plant, or is converted to heat. It is estimated that of the light energy impinging on the surface of vegetation, on average as little as 1–5 per cent, may be trapped as food energy. Yet despite this low figure, enough energy is trapped to maintain all life.

(b) The use of light energy to manufacture carbohydrates in photosynthesis involves many complex chemical processes, each catalysed by enzymes. The overall reaction can be summarised as follows.

$$6CO_2 + 6H_2O + \text{LIGHT ENERGY} \rightarrow 6C_6H_{12}O_6 + 6O_2$$

from the air or from respiration	from the soil	absorbed by the pigments in the leaf (mainly chlorophyll)	sugar in plant cell	released to in the air or used in respiration

(c) The sugar produced in photosynthesis has several potential destinations. It can be converted into relatively stable energy-rich substances, such as starch, and be stored. It can be combined with other sugar molecules to form specialised carbohydrates, such as cellulose, or it can be combined with other substances and nutrient compounds to build the complex molecules of proteins, pigments and hormones. All of these transformations and reactions require energy from respiration.

5. The process of respiration. Respiration is basically the reverse of photosynthesis. It involves complex reactions and uses many of the same enzymes as photosynthesis. The overall reaction can be summarised as follows.

$$6C_6H_{12}O_6 + 6O_2 \rightarrow 6CO_2 + \text{USABLE ENERGY}$$

produced by photosynthesis	from the air or from photosynthesis	released to the air or used in photosynthesis	lost from the ecosystem

In reasonable conditions photosynthesis proceeds up to 30

times faster than respiration, but it takes place only in the light. In all plants a great deal of energy is used up by respiration. The actual proportion of carbohydrates used up in this way varies from about 10 to 75 per cent depending on many variables including the species and age of the plant.

FACTORS CONTROLLING THE RATE OF PHOTOSYNTHESIS

The speed and efficiency of photosynthesis depend on a number of factors, some of which are environmental and some of which are inherent characteristics of different species of plant. In 1905, F.F. Blackman realised that when a process is influenced by a number of factors, its rate will be governed by the factor in least supply. He called this the "law of limiting factors". It is particularly relevant to photosynthesis which may be limited by a number of different variables.

6. Light. Both the quality (wavelength) and the intensity of light are important.

(*a*) The light energy is absorbed by the pigments of the plant. The most important of these is chlorophyll which is green and which therefore absorbs light of the red and blue wavelengths. Green light is reflected by chlorophyll and cannot be used by plants unless they have additional pigments (usually brown or blue as in some seaweeds) which can absorb light of the green wavelengths. Consequently, the quality or wavelength of light reaching the plant may be a limit on photosynthesis.

(*b*) The intensity of light governs the actual amount of energy impinging on the surface of the plant and hence the amount of energy available for photosynthesis. Laboratory studies suggest that in conditions of dim light, plants can convert absorbed solar energy to sugar with an efficiency of about 20 per cent. In bright light this efficiency decreases progressively to a plateau level at about 8 per cent. This pattern is true for all photosynthetic plants ranging from minute algae to large trees and seems to be due to the inability of the photosynthetic mechanism to keep pace with the increased input of energy. Supplies of other essentials, such as carbon dioxide, may limit the process even if light is abundant.

If the intensity of light is very great it can damage the plant by breaking down the chlorophyll.

(c) If all other requirements are satisfied, the amount of light energy a plant can use is determined by the amount it can absorb, i.e. by the amount of chlorophyll present. Certain plants, such as beans and cereals, are adapted to live in habitats with high light intensities and contain a high proportion of photosynthetic tissue in their leaves. Other plants, such as wood sorrel, are adapted to live in low light intensities. Their leaves contain smaller amounts of photosynthetic tissue and therefore have lower total amounts of chlorophyll in them. If exposed to high light intensities, the chlorophyll of this type of plant becomes saturated and cannot absorb all the light available.

7. Carbon dioxide. This diffuses into the plant from the atmosphere. In most plants this occurs through the pores in the leaf known as *stomata* (singular = *stoma*, from the Greek meaning mouth), which are usually open in the daytime and closed at night. The uptake of carbon dioxide is a passive process but is influenced by several factors, the most important of which are the relative magnitudes of the concentrations inside and outside the plant, and the opening of the stomata. It is thought that in many cases insufficient carbon dioxide diffuses into the plant to utilise all the energy absorbed by the chlorophyll. In this way carbon dioxide supplies often limit photosynthesis.

8. Water. If the plant does not have adequate supplies of water its stomata will close and the plant will start to wilt. In this condition all the metabolic processes of the plant, including its photosynthesis, will be slowed down.

9. Nutrients. In order to photosynthesise efficiently, a plant must have sufficient supplies of all the nutrients required to manufacture the chlorophyll and enzymes involved in the process. Many of the trace elements or micronutrients are important in this respect. For example, magnesium forms a vital part of the chlorophyll molecule.

10. Temperature. The rate of all chemical processes is governed by temperature. The Hoff rule states that for every 10 °C (18°F) increase in temperature, the rate of a chemical process will double. This is only partly true for organic chemical processes as the rate of these generally increases to a maximum at an optimum temperature then declines quickly thereafter. At high temperatures the enzymes aiding the reaction become deactivated.

THE INFLUENCE OF COMMUNITY STRUCTURE
AND COMPOSITION ON PRIMARY PRODUCTIVITY

Primary productivity may vary markedly even between eco-systems functioning in identical conditions of climate and soil. This is mainly the result of contrasts in their community composition and structure.

11. Species and age of plant. There may be highly dissimilar growth rates among the different competing species within the same ecosystem. In addition productivity varies with age. Most plants achieve their maximum net productivity when they are young. It is an ecological advantage to grow quickly at first in order to compete effectively. For example, barley can fix about 14 per cent of total incident light energy as food during the first few months of growth. These youthful high productivity rates are maintained for short periods only. As the size of the plant body increases, more energy is required to maintain it. The percentage of gross primary productivity left as surplus for net productivity gradually decreases as the plant matures (*see* Fig. 11).

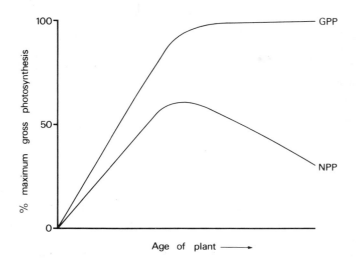

FIG. 11 *General changes in the pattern of productivity in a plant as it ages. GPP: Gross primary productivity; NPP: net primary productivity.*

12. Shading. The geometric form of the plants and the density of their spacing will have a great influence on the efficiency of the ecosystem. For instance, maize and sugar cane are among the most efficient and productive crop plants. These have relatively broad, vertical leaves to provide the maximum photosynthetic area to sunlight with the minimum amount of mutual shading.

In complex natural ecosystems, like forests, there may be several layers of vegetation including a tree canopy, a shrub layer and a layer of short plants on the forest floor. The amount of light available for these understorey layers will be determined by the characteristics of the canopy layer such as the shape of the trees, their height and their reflectivity.

BASIC PROCESSES INVOLVED IN SECONDARY PRODUCTIVITY

Gross secondary productivity is the total amount of organic material assimilated by animals, that is the organic material which actually passes through the gut wall:

gross secondary productivity = total amount of material ingested
− amount of material lost by defecation

As in the case of plants, the food assimilated by animals can be stored as carbohydrates, proteins or fats, transformed into relatively simple substances or rebuilt by the animal into more complex organic molecules. However, since animals cannot manufacture their own food, they face the problem of converting ingested materials to those suitable for use in their own bodies. This is particularly relevant in the case of the herbivores and gives rise to an essential difference between their nutrition and that of the carnivores.

13. Herbivore nutrition. Food digestion is harder for herbivores than for carnivores, not only because they must change plant materials into animal materials but also because they have to break down the cellulose cell walls of the plants before the cell contents can be digested.

Cellulose is a very resistant substance which only a few animals, such as the wood-boring beetles, can digest on their own. Most herbivores rely on the activity of micro-organisms (mainly anaerobic bacteria and protozoans) in their guts to break down the cellulose into small usable molecules for them. This activity takes some time to operate so food is generally held in the bodies of herbivores longer than it is in those of carnivores. The herbivores can be divided into two groups on the basis of the location of the cellulose digestion in the gut.

(*a*) *Ruminants*. These animals, which include cattle, sheep, antelopes and camels, have their cellulose-digesting micro-organisms in the stomach. The stomachs of ruminants are often divided into four chambers, the largest of which is called the *rumen*. This acts as a fermentation chamber for the digestion of cellulose.

(*b*) *Non-ruminants*. These herbivores, for example rats, mice, rabbits and squirrels, have their cellulose digesting micro-organisms in a small sac called the *caecum*, which is at the junction of the small and large intestines. This is a less efficient location than the stomach because it occurs late in the gut and therefore the products of digestion have less chance of being absorbed.

In addition, the non-ruminants generally possess fewer cellulose-digesting micro-organisms than do the ruminants. Some animals have adapted to this problem by passing food through the gut twice, a process known as *refection*. For example, the food ingested by rabbits is only partially processed during its first passage through the gut. It is expelled through the anus as pellets which are then re-eaten to complete digestion.

14. Carnivore nutrition. Carnivores are not able to digest cellulose at all. The food they obtain from the bodies of herbivores has already been converted to animal materials and generally has a higher calorific value than vegetation. As organic material is passed up the trophic levels it is converted to higher energy food molecules.

15. Heterotroph respiration. The energy required to perform all the necessary metabolic processes of the herbivores and the carnivores is provided by respiration. In all animals the efficiency of energy conversion is low so a great deal of energy is lost from the ecosystem at each trophic level. Respiration losses account for a significant proportion of gross secondary productivity throughout ecosystems, but actual amounts do vary according to the position of the animal on the food chain. Animals at high trophic levels, such as wolves, must expend relatively more energy in moving about to find prey, than animals at low trophic levels, such as rabbits, do in eating their food.

ECOLOGICAL EFFICIENCY

The efficiency of energy transfer from one trophic level to the next is the ecological efficiency of the ecosystem. There are several possible methods of expressing this but two ways are used most frequently.

16. Gross ecological efficiency. This is the ratio of the gross productivity of any trophic level to that of the gross productivity of the trophic level preceding it:

$$\text{gross ecological efficiency} = \frac{C_p}{C_e} \times 100\%$$

where C_p refers to the calories of prey eaten by a predator, and C_e refers to the calories of food eaten by the prey.

17. Food chain efficiency. This is the ratio of the gross productivity of any trophic level to that of the energy supplied to the preceding trophic level. It is essentially a refinement of the gross ecological efficiency measure, and takes account of the fact that at each trophic level a lot of food usually passes straight to the decomposers:

$$\text{food chain efficiency} = \frac{C_p}{C_s} + 100\%$$

Where C_p refers to the calories of prey eaten by a predator, and C_s refers to the calories of food supplied to the prey but not necessarily eaten by it.

18. Measurements of ecological efficiency. Attempts have been made to measure ecological efficiencies both in laboratory and in natural systems. In 1959, Slobodkin examined the interactions of a simple aquatic food chain composed of:

producer   = *Chlamydomonas* (unicellular alga)
      ↓   ↓
herbivore = *Daphnia*   (water flea)
      ↓   ↓
carnivore = Man   (acting as a carnivore by removing *Daphnia* from the system)

He found that the maximum gross ecological efficiency which the system could achieve was about 13 per cent. The rest of the energy was dissipated during transfer. Slobodkin found that if the system was altered to increase the rate of "predation" so that the herbivores were overcropped, the gross ecological efficiency decreased to less than 25 per cent of the potential because the energy from the producers was wasted.

The few studies which have been conducted on natural systems indicate that gross ecological efficiencies rarely exceed 10 per cent. This is very significant as it means that in a typical ecosystem one can expect about 90 per cent energy loss between the trophic levels of a grazing food chain. Even at their most productive young stages of growth, animals rarely exceed 35 per cent efficiency of energy transfer. At first it may seem strange that the efficiency of productivity in ecosystems should be so low, but it must be remembered that a lot of evolutionary pressures are exerted in the system. Selective forces for breeding, escaping from predators or maintaining territory may take precedence over the importance of ecological efficiency.

19. Implications for food chain length. We have already noted that the number of steps on a food chain is limited by the dissipation of energy both within and between trophic levels. The existence of a species at a particular trophic level depends not only on its capacity to assimilate sufficient energy to compensate for its losses to respiration and predation but also on having enough net productivity to allow for growth.

Since the gross ecological efficiency of most ecosystems is only 10 per cent, the energy available to organisms decreases exponentially through the food chains (*see* Fig. 12). If the energy required by organisms at each trophic level decreased in a similar pattern, the food chains could have an indefinite number of links. In fact, the opposite is true. Organisms at high trophic levels tend to dissipate more energy in respiration that do those at the lower trophic levels.

The figures given in Table I illustrate that the energy requirements for organisms generally increase with progressive steps on food chains. Consequently, the net productivity of high trophic levels is small and the number of steps on food chains is restricted. A limit to the chain will be imposed when organisms cannot achieve any net productivity. In most small ecosystems (e.g. Cedar Lake Bog, Table I) this occurs at the primary carnivore level, but even in large productive systems there are rarely more than five trophic levels.

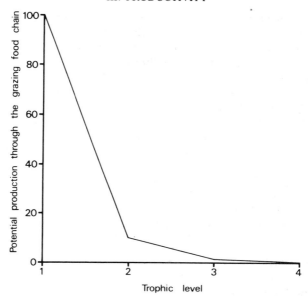

FIG. 12 *Relative potential production through the grazing food chain taking autotroph productivity equal to 100 units.*

TABLE I. EXAMPLES OF RESPIRATORY LOSSES AT SEVERAL
TROPHIC LEVELS IN DIFFERENT ECOSYSTEMS

| Ecosystem | *Percentage respiratory losses* | | | |
	Producers	Herbivores	Primary carnivores	Secondary carnivores
Lake Mendota, Wisconsin[1]	22.3	36.1	47.8	66.7
Cedar Lake Bog, Minnesota[2]	25.0	38.1	58.1	–
Silver Springs, Florida[3]	57.5	56.1	82.5	61.9
Salt Marsh, South Carolina[4]	77.5	77.7	81.3	–

Figures given as percentage differences between gross and net productivity. Data from: Lindemann 1942[1,2]; Odum 1957[3]; Teal 1962[4].

THE ECOLOGICAL EFFICIENCIES OF GRAZING
AND DETRITUS FOOD CHAINS

20. Comparative efficiencies. The grazing and detritus food chains contrast markedly in their ecological efficiencies. Consequently, it is important to determine the proportion of energy flow through each of them when considering the functioning of an individual ecosystem. In most circumstances the detrital food chain is far less efficient than the grazing food chain.

Decomposer organisms tend to be very active, generate a lot of heat in their respiration and so liberate a great deal of energy from the ecosystem. In many terrestial ecosystems, such as forest (*see* Fig. 13), the detrital food chain plays an important role in energy transfer through trophic levels. In cases like this, losses of energy from the detrital food chain, account for a large proportion of the total energy lost from the ecosystem.

In contrast, in the majority of marine ecosystems, such as the one depicted in Fig. 13, the detrital food chains convey a smaller percentage of the total energy flow. However, respiratory losses are still large compared with those from the grazing food chain.

21. Anaerobic and aerobic detrital food chains. The ecological efficiency of the detrital food chain depends on many factors such as temperature and acidity, but most importantly it is influenced by the oxygen content of the detritus. In aerobic conditions the decomposers use the normal process of respiration to release energy. However, oxygen is lacking in many places where detritus accumulates (e.g. the bottom of ponds). In these situations the breakdown of organic compounds and the release of energy must proceed anaerobically.

One of the most important methods involved in this is fermentation, which splits large molecules into smaller ones without the addition of oxygen. The release of energy is much less than that from aerobic respiration, but it is still sufficient for the metabolism of the decomposers.

Fermentation results in the partial breakdown of organic material which leads to an accumulation of undegraded detritus. This means that the gross ecological efficiency of the anaerobic detrital food chain is less than that of the aerobic one because much of the food energy is not passed to subsequent trophic levels.

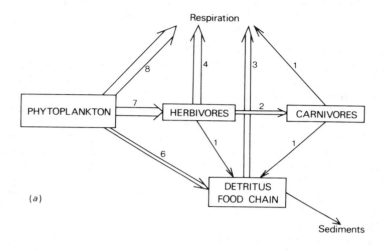

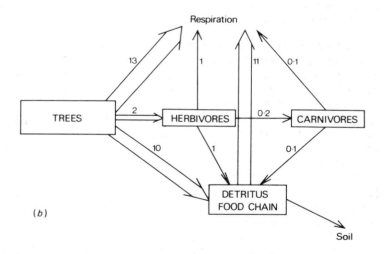

FIG. 13 *Energy flow through grazing and detritus food chains. (a) Marine chains; (b) forest community. (Modified from Odum, 1971).*

METHODS OF MEASURING PRIMARY PRODUCTIVITY

It is necessary to pay some attention to how primary productivity is measured since this process is of great ecological significance. A knowledge of measurement techniques gives insight into the problems involved in comparing figures derived from different ecosystems, and helps to explain why very little quantitative work has been done on productivity. Most measurements are based on some indirect quantity such as the amount of a substance produced, the amount of raw material used up or the amount of byproduct released. It must be remembered that in all of these, the processes and products of photosynthesis are balanced in part by those of respiration.

22. Harvest method.

(a) In this the weight of the growth produced by the plants is determined. It can be expressed either directly as dry weight or indirectly as calorific value, but in both cases it is measured per unit area per unit time. Since the actual weight of growth is determined, the harvest method is a measure of net primary productivity.

(b) The method is most suitable for terrestrial ecosystems which have no grazers, e.g. crops of annual plants. In these, weight of growth increases from zero to a maximum at the time of harvest. The technique can be applied also to natural ecosystems in which annual plants predominate and in which there are few grazers. In these circumstances it is best to sample productivity at intervals on a series of plots through the growing season, rather than to take one estimate at the end of the season. This is because plants in natural ecosystems will have different life-spans, so some may die off before the end of the season and be missed from the productivity weight.

The harvest method is simple but has several sources of potential error. For example, all of the root systems should be included in the weight of growth. Also, practically all ecosystems have some herbivores even if they are only inconspicuous insects feeding on the crops.

23. Oxygen method (light and dark bottle method).

(a) Oxygen is a byproduct of photosynthesis so there is a definite relationship between productivity and the amount of oxygen given off from a plant. However, some of the oxygen may be used up in respiration. This has to be taken into account in the calculations. Oxygen measurement is most suitable for use in aquatic ecosystems where algae are the main producers.

(b) Two equal samples of water (containing the algae) are taken from the same depth in the ecosystem. One sample is put in a transparent (light) bottle and the other is put in a dark bottle. The oxygen content of both samples is determined then the bottles are shut and returned to their original depth in the water.

(c) The two bottles are left in position for between 1 and 12 hours. During this time, the oxygen content of both bottles will alter. In the dark bottle respiration will take place, using up oxygen. In the light bottle, both respiration and photosynthesis will occur. Since photosynthesis normally proceeds more rapidly than respiration it will produce more oxygen than respiration uses up. The amount of respiration in the two bottles is assumed to be equal.

(d) At the end of the exposure time the oxygen content of both bottles is measured again. The amount of oxygen produced by photosynthesis is the amount produced in the light bottle plus the amount used up by respiration in the dark bottle. Once the amount of oxygen produced has been determined, the primary productivity of the algae can be estimated.

(e) This technique is widely used in fresh water and marine ecosystems but has several disadvantages. For instance, it can be applied to small producers only. In addition, it assumes that respiration proceeds at the same rate in the light as in the dark. This may not be so because respiration shares many catalytic enzymes with photosynthesis and hence the speeds of the two processes may be related.

24. Carbon dioxide methods. Carbon dioxide is used up in photosynthesis and so the amount absorbed by plants can be taken as an indication of primary productivity. As in the case of oxygen, the influence of respiration must be considered.

The measurement of carbon dioxide is suitable for use in terrestrial ecosystems. The methods can be applied to single leaves, whole plants or even entire communities. Two main techniques are used.

(a) The enclosure method is suitable for use with parts of plants or with whole plants of small sizes. Two sample plots are selected. These should be as similar as possible. One plot is enclosed in a light container, the other in a dark container. Unlike the oxygen method, air is allowed to enter and leave the containers in pipes.

The concentrations of carbon dioxide in the air passing to and from each container are monitored. In this way, the amount of carbon dioxide used up by photosynthesis can be determined. It will be equal to the amount produced in the dark container (carbon dioxide from respiration) added to the amount used up in the light container (carbon dioxide taken in for photosynthesis but produced from respiration).

This method not only suffers from the same disadvantages as the light and dark bottle oxygen method but also, on land, the enclosures tend to act like greenhouses. The sample plots warm up, so the rates of photosynthesis and respiration will change.

(*b*) The aerodynamic method of carbon dioxide measurement avoids some of the disadvantages of the enclosure method by monitoring without containers. Carbon dioxide measurements are taken by sensors mounted on a vertical mast erected in the community so that its top is taller than the plants.

Changes in the concentrations of carbon dioxide above and within the community can be taken as an indication of productivity. At night, the carbon dioxide concentration within the community will increase due to respiration, whereas in the daytime it will decrease because of photosynthesis. Comparisons of the concentrations will indicate how much carbon dioxide has been used for photosynthesis.

The aerodynamic method seems to have great potential for future development but suffers from the major disadvantages that it can be used only when there is little air movement. Wind will blow away the carbon dioxide and distort the readings.

25. The acidity method. In aquatic ecosystems the acidity of the water will be a function of its dissolved carbon dioxide content. Changes in acidity can be used as an index of productivity. The relationship between acidity and carbon dioxide content is complex and so must be interpreted individually for each ecosystem examined.

This method is subject to a lot of potential errors because many other factors, such as nutrient content, effect the acidity of the water.

26. The disappearance of raw materials. Productivity can be measured by the rate of disappearance of raw materials such as nitrates and phosphates. The method measures the net primary productivity of the ecosystem. This technique is most useful for large aquatic ecosystems such as lakes and oceans where nutrients accumulate through the winter and are then used up in the spring.

The main disadvantage of the method is that the uptake of raw materials may be compensated by return via the nutrient cycles.

27. Radioactive tracers. A known amount of "marked" material which can be identified by its radiations is introduced to the ecosystem. For example, radioactive carbon [^{14}C] could be introduced in the carbon dioxide supply. Its assimilation by plants is monitored to give an accurate assessment of productivity.

This technique is expensive and requires elaborate equipment but its main advantage is that it can be done in a wide variety of ecosystems and causes less disturbance to the system than other methods.

28. Chlorophyll method. Productivity is related to the amount of chlorophyll present. The assimilation ratio for a plant or an ecosystem is the rate of productivity per gram of chlorophyll. The concentration of chlorophyll can be determined by relatively simple procedures involving the extraction of pigments from the plants. Generally, the total chlorophyll per unit area is higher in terrestrial ecosystems than in aquatic areas.

If the assimilation ratio, chlorophyll concentration and the amount of light energy reaching the plants are known, then gross primary productivity can be estimated.

This technique can be applied to a wide variety of ecosystems.

PRIMARY PRODUCTIVITY IN NATURAL ECOSYSTEMS

29. Causes of variation in primary productivity. Measurements of productivity in natural ecosystems have revealed a wide range of amounts both spatially and temporally. These differences can be attributed to three main causes.

(*a*) *Intensity and duration of sunlight.* The annual input of energy from the sun varies globally from a maximum in the tropics to a minimum at the poles. Consequently, the greatest potential for annual productivity occurs in the tropics and decreases with increasing latitude. At a seasonal level, the low energy input of temperate regions is compensated partly by the long days of summer. If summer months only are considered, the energy input to temperate latitudes is quite high.

(*b*) *Environmental limitations.* Even if sunlight is abundant, primary productivity may be limited by environmental factors like lack of water, low temperatures and deficiences of nutrients. These environmental controls on distribution and function are considered in detail in V and VI but it can be stated that, usually, warmer climates are more productive than cold ones, and wet climates are more productive than dry ones.

(*c*) *Community composition and structure.* As we have noted earlier in this chapter (*see* **11** and **12**) factors such as species present, shape of plant and spacing, have a great influence on primary productivity.

30. Estimates of productivity in natural ecosystems. Almost all measurements in natural ecosystems have to date revealed very low rates of gross primary productivity. In 1963 E.P. Odum, an American ecologist, attempted to produce a generalised survey of gross primary productivity in selected world ecosystems. His results, which are accepted widely, revealed that some ecosystems were relatively unproductive while others were surprisingly fertile. Odum's measurements, shown in Fig. 14, represent average gross primary productivity in terms of g/m²/day of dry organic matter over an annual cycle. They can be grouped into three orders of magnitude.

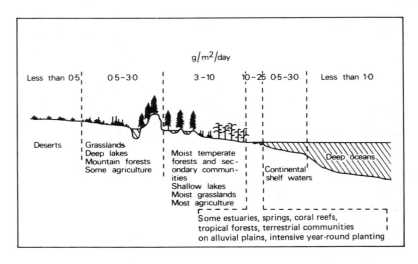

FIG. 14 *The world distribution of primary production. (After Odum, 1971).*

(a) *Relatively unproductive*. These include some parts of the open ocean and the deserts. Productivity is below 0.1 g/m^2/day.

(b) *Moderately productive*. These include semi-arid grasslands, coastal seas, shallow lakes and dry forests. Productivity is between 1 and 10 g/m^2/day.

(c) *Very productive*. These include estuaries, coral reefs, moist forest, alluvial plains and intensive agriculture. Productivity is between 10 and 20 g/m^2/day.

Odum found that productivity rates in excess of 20 g/m^2/day occurred occasionally for limited periods in some experimental crops and polluted waters.

31. General statements about the productivity of ecosystems. Certain general patterns in the world distribution of primary productivity emerge from the data produced by Odum and other workers.

(a) A very large percentage of the earth's surface is in the low productivity category, either because of lack of water (deserts) or lack of nutrients (sea).

(b) The productivity of the oceans is appreciably lower than that of the land despite its greater area and prolific life. This difference derives from several causes. The most important of these are namely the higher percentage of light absorbed and reflected by water than by the atmosphere, the high proportion of energy used up in respiration by the phytoplankton, and lastly nutrient deficiency, especially in the surface layers.

(c) The most productive ecosystems are very open systems that have extensive communications with and inputs from other ecosystems. For example, estuaries, swamps and coral reefs all receive large inputs of nutrients from adjacent systems. Semi-closed systems which have more self-contained nutrient cycles are generally less productive.

(d) There seems to be a definite upper limit to the efficiency with which light energy may be harnessed for productivity on any large scale. This maximum has already been attained by some natural ecosystems such as coral reefs.

PRODUCTIVITY IN AGRICULTURAL SYSTEMS

32. Estimates of productivity.

(a) The average utilisation of solar energy by natural

ecosystems is between two and seven times that of the average crop. This has serious implications because clearly most of our current patterns of food production are inefficient.

Measurements of actual net primary productivity made by the International Biological Programme in 1975 showed that few natural ecosystems even approached the theoretical limits for productivity. If the natural systems are replaced by agriculture, efficiency usually decreases. For example, world average grain productivity is about 2 g/m^2/day which is a low figure compared with the productivity of most natural systems.

Actual net productivity figures approach the theoretical limits only in certain tropical agricultural systems, such as sugar cane, and in some intensive temperate cultivation (*see* Table II). On a year-round basis, the average efficiency of gross productivity in tropical agriculture, without irrigation, is between 0.5 and 1 per cent.

(*b*) In any climate area the agricultural systems which utilise solar energy most fully are those which have a year-round cultivation with a dense crop canopy. For example, in the European mediterranean the mixed tree and herb system of olives with wheat and barley beneath is very efficient. In contrast, the majority of western cultivation techniques tend to leave large open spaces between individual plants. This decreases the amount

TABLE II. RELATIONSHIPS BETWEEN GROSS AND NET
PRODUCTIVITY IN CROPS UNDER INTENSIVE CULTIVATION

| | cal/m^2/day | | | % | |
	SR	GP	NP	GP/SR	NP/GP
Sugar cane, Hawaii	4000	306	190	7.6	62
Irrigated maize, Israel	6000	405	190	6.8	47
Sugar beet, England	2650	202	144	7.7	72

SR: solar radiation; GP: gross productivity; NP: net productivity. Figures are for favourable growing season conditions. Data from Monteith, 1965.

of chlorophyll exposed per unit area and so less light energy is used.

(c) Practically all productivity estimates for agriculture conceal energy subsidies. Cultivation requires the expenditure of additional energy in forms such as fuel for tractors and power to make fertilisers and insecticides. If these are taken into account, agriculture is found to be even less efficient.

33. Implications for human nutrition.

(a) Since the actual productivity of both natural and agricultural systems is low, there would appear to be great potential for improving efficiency. The ways in which this could be done include minimising limiting factors (e.g. irrigating when water is limiting productivity), breeding crops with higher yields, and improving agricultural techniques such as by harnessing the maximum amounts of solar energy per unit area whilst avoiding excessive mutual shading of plants.

Most authorities doubt whether these combined methods could achieve more than a ten-fold increase in food production. This would be roughly equivalent to the present rate of population increase.

(b) The inefficiency of primary productivity in agriculture is aggravated by the fact that large numbers of people occupy secondary trophic levels. We have seen that a high percentage of energy flow is dissipated through the food chain. Far more people can be kept alive on an acre of wheat than on the animals living off an acre of alfalfa. Although the net secondary productivity of young cattle may be as high as 35 per cent this declines to about 11 per cent when the animals mature.

Increasing pressure on world food supplies may force man to shorten his food chains and utilise more available space for edible plants.

PROGRESS TEST 3

1. Explain the term gross primary productivity. **(2)**
2. What is the basic reaction in photosynthesis? **(4)**
3. Which factors influence the rate of photosynthesis? **(6–10)**
4. Why is food digestion harder for herbivores than it is for carnivores? **(13)**

5. What is meant by food chain efficiency?   (**17**)

6. How are food chains limited by ecological efficiency? (**19**)

7. Why are detrital food chains less efficient than grazing food chains?   (**20**)

8. Explain the "light and dark bottle" method of measuring primary productivity.   (**23**)

9. What is the assimilation ratio of a plant?   (**28**)

10. Which variables control primary productivity in natural ecosystems?   (**29**)

11. Why are marine ecosystems unproductive?   (**31**)

12. Why is primary productivity low in agricultural systems? (**32**)

CHAPTER IV

Ecological Succession

INTRODUCTION

1. Causes of change in ecosystems. All ecosystems undergo changes in their structure and function through time. Some of these changes are minor local fluctuations and are relatively unimportant. Others bring about major alterations in the assembly of species present and effect the stability of the system as a whole. The study of ecosystem change and stability has received a great deal of attention because it has wide-ranging influences on many aspects including nutrient cycling, productivity, agriculture and conservation. Major changes in ecosystems derive from three main causes.

(*a*) *Changes in climate*. Oscillations in the climates of the world over thousands of years have brought about corresponding adjustments in the world's ecosystems. This type of change has involved long-term alterations in the distributions of plants and animals, and hence the different assemblies which can come together to form ecosystems.

(*b*) *The influence of local external factors such as fire, trampling or pollution*. These may induce changes in ecosystem structure and function either on a short-term or a long-term basis.

(*c*) *Development due to the inherent nature of the system*. This is ecological succession which can be defined as the changes which take place in an ecosystem as it develops towards maturity or steady-state. We have noted in I that ecosystems are open systems and have the capacity for self-regulation by negative feedback. This means that they will tend towards a state of equilibrium. This type of change is by far the most important in terms of ecosystem stability.

2. The basic idea of ecological succession. It is common knowledge that if a garden is left untended or if a pasture is not grazed, the vegetation will not stay the same indefinitely. Weeds and other plants will invade and will alter the character of the community. If agricultural land is abandoned, shrubs and trees will eventually become established on it, provided that conditions of soil and climate are suitable, so the area will develop a forest cover.

Similar changes occur on completely new areas of land such as deltas, sand-dunes and newly formed volcanoes. Places like these have not supported life before. They have no soil or organic nutrients; their surfaces are exposed so the conditions for life on them are extremely harsh and adverse. However, given sufficient time, the areas will become colonised by plants and animals and ecosystems will develop.

3. An example of succession. On volcanic lava the first plants to grow are usually lichens and mosses which can withstand exposure and desiccation. These pioneer plants start to break up the surface of the lava, so forming a simple soil. When they die their bodies provide a food source for pioneer decomposers. The organic matter from them accumulates and provides raw materials for other plants to use.

Gradually, the pioneer plants ameliorate the conditions for life. Hardy grasses and other herbaceous (non-woody) plants colonise and continue the process of soil formation by breaking up the surface of the lava with their roots and by liberating organic matter when their bodies decompose. The grasses and herbaceous plants tend to shade out the lichens and mosses which first colonised the area.

However, over time, as the soil develops and as the conditions for life become less harsh, shrubs will invade and shade out many of the grasses and herbaceous plants. Eventually, trees will be able to grow in the area and will replace many of the shrubs. When the trees have become established there will probably be little further change in the type of vegetation present. The ecosystem will have reached a state of equilibrium.

The development and changes in the vegetation communities will be accompanied by parallel changes in animal communities. Different types of animal will be able to tolerate the conditions in the ecosystem at the various stages of its development.

In succession the changes take place because of the inherent nature of the ecosystem. The development of a mature ecosystem by ecological succession involves a sequence of communities of different species. Each community ameliorates the environmental conditions slightly, so preparing the habitat for colonisation by the next community until finally there is little more change as the ecosystem becomes mature.

4. Definitions of basic terms.

(*a*) *Primary succession*. This is the process of ecological succession in places which have not supported ecosystems before, for example, sand-dunes, deltas, volcanoes, glacier moraines and new ponds. All these areas are biotically inactive at first.

(*b*) *Secondary succession*. This is the process of ecological succession in areas which already support ecosystems but in which development to maturity or steady-state has been prevented by an arresting factor such as grazing or fire. If the arresting factor is removed, secondary succession will take place to develop the ecosystem to maturity.

(*c*) *Sere*. The sequence of communities in a succession is known as a sere. Different terms can be applied to describe the conditions under which the succession takes place. For example, succession starting in very dry conditions is called a *xerosere* and that starting in water is called a *hydrosere*.

(*d*) *Seral stage*. The individual communities of the succession are known as seral stages. The seral stages may not be distinct but rather tend to merge into one another as the ecosystem undergoes continuous change.

(*e*) *Climax community*. Traditionally this term is applied to the community which is established at the end of a succession and which undergoes little further change. It can be regarded as the ultimate development of the ecosystem.

TRADITIONAL APPROACHES TO SUCCESSION

5. The historical development of the idea. The theory of ecological succession was first proposed by H.S. Cowles of Chicago University in 1899. His careful observations of the sand dunes round Lake Michigan led him to infer that the bare, newly formed dunes close to the water's edge would gradually become colonised by vegetation. Cowles concluded that as the dunes aged the communities which they supported changed. Initially the vegetation was dominated by dune grasses but these were invaded by cottonwood, which was in turn taken over by pine and oak, until eventually a mature beech and maple forest developed as the climax of the sere.

Cowles could not actually observe this pattern of succession taking place through time, because the changes took a long time to occur, but he deduced the sequence of communities which he thought would develop by examining dunes of different ages.

His beliefs were reinforced by those of his student, F.E. Clements, a respected traveller and botanist. Clements had observed secondary successions taking place on the American prairies. He noticed that vegetation tended to restore itself whenever it was disturbed. For example, it grew over the migratory trails of the vanished buffalo and over the ruts left by wagon trains.

In 1916, Clements published his ideas in a very persuasive manner, stating that vegetation could be thought of as a "superorganism" which tried to repair itself whenever it was damaged. He thought that if natural communities were removed, as for instance to clear land for agriculture, the vegetation would try to re-establish itself if the land was abandoned. Clements convinced the majority of American botanists to look at plant communities in terms of seral stages striving to develop towards climax communities. Cowles and Clements together worked out a logical and academically attractive theory of succession which became the basis of ecological thought for many decades.

6. The main points of the traditional theory of succession.

(a) Succession is an orderly process of community development that involves changes in species composition and ecosystem function through time. It is a progressive process and is therefore predictable.

(b) Early seres are simple in structure, and are dominated by shortlived plants. Later seres become progressively more complex and are dominated by longlived plants.

(c) Succession culminates in the climax community which is the largest, most efficient and most complex community which the habitat can support. The climax is a stable, self-perpetuating community.

(d) Successions in habitats which are different initially can lead to the same climax communities. This idea is known as *equifinality*. For example, in the British Isles hydroseres and xeroseres can both result in the development of a climax oak or beech forest (*see* Fig. 15).

(e) The most important factor influencing the nature of the climax community is climate. Cowles and Clements thought that for each climatic area there would be one type of climax community. This is the *monoclimax theory*. Local variations in climax communities, such as those due to soil and drainage, were

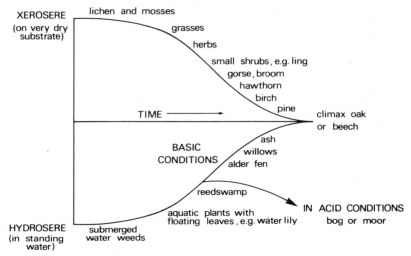

FIG. 15   *The idea of equifinality illustrated by xerosere and hydrosere development in the British Isles. (After Ashby, 1961).*

accounted for by describing them as temporary phenomena. Clements thought that, given sufficient time, these local communities would develop into the regional climax type.

7. The need to reassess the traditional approach. The traditional theory of succession is very rigid. It relies heavily on deductive thinking and circumstantial evidence. Very few cases of succession have been studied in detail because the changes take place over decades and it is difficult to maintain field measurements over these time-spans.

In view of this, it is surprising that the traditional concept of succession has dominated the development of ecology this century. Many research workers have devoted themselves to forming general theories about succession and to searching for unifying themes. In recent years many of the views which were widely accepted by the followers of Cowles and Clements, have been challenged. It is here, probably more than in any other branch of ecology that ideas are undergoing radical change.

In order to put both the traditional and the latest theories into perspective, it is necessary to consider all the aspects of ecosystem structure and function which change during succession. These are dealt with in detail in the rest of this chapter and for convenience are summarised in Table III.

TABLE III. A SUMMARY OF THE MAIN CHANGES THAT OCCUR IN AN ECOSYSTEM DURING SUCCESSION

Feature	Development seres	Mature or climax stage
SPECIES STRUCTURE		
Species composition	Rapid changes	Gradual changes
Species diversity	Increases	Stabilises or declines
LIFESTYLES		
Size of dominant plants	Small	Large in terrestrial systems May be small in aquatic systems
Lifecycles	Short, simple	Long, complex in terrestrial systems May be short, simple in aquatic systems
Strategies	Generalists	Specialists
ORGANIC STRUCTURE		
Total biomass	Increases	Maximum amount present
Stratification	Simple	Complex
ENERGY FLOW		
Trophic relationships	Short, linear food chains	Long, complex food webs
Gross productivity	Low	High
Net productivity as percentage of gross	High	Low
Ecological efficiency	Increases towards mid-succession	Declines
Stability (resistance to disruption)	Low	High
NUTRIENT CYCLING		
Nutrient cycles	Open	Closed
Amounts cycled	Small	Large in terrestrial systems May be small in aquatic systems
Exchange rate between abiotic and biotic components	Fast	Slow
Role of detritus in regenerating nutrients	Unimportant	Important

CHANGES IN ENERGY FLOW PATTERNS
AND PRODUCTIVITY

8. Energy flow patterns. During succession to climax the pattern of energy flow in an ecosystem changes fundamentally. The changes are reflected in the amount of standing crop in the system.

(*a*) During the early seral stages energy inputs to the system are greater than the losses (*see* Fig. 16(*a*)). Plant and animal communities develop, accumulating energy as biomass. The amount of standing crop present in the ecosystem increases during succession.

(*b*) When the climax community has developed a steady state is attained. In this, energy inputs to the ecosystem equal energy losses (*see* Fig. 16(*b*)). As a result, there is little further change in the standing crop. The flow of energy through the system is at its maximum at the climax stage.

(*c*) If the ecosystem is disturbed by some external factor, such as fire, energy losses may exceed energy inputs (*see* Fig. 16(*c*)). In this case, the amount of standing crop present in the system will decrease.

(*d*) The accumulation of energy as biomass during succession is most marked in terrestrial ecosystems. In these, the largest plants of the succession form the climax community (usually trees). The standing crop is at its maximum at this stage although the actual amount present may fluctuate slightly (*see* Fig. 17).

(*e*) In aquatic ecosystems, especially marine ones, the climax or steady-state community may be represented by phytoplankton. The small size of these plants means that the standing crop is relatively small and it may seem that there is little accumulation of energy in the system. However, the high metabolic rate of phytoplankton enables them to have high gross productivity, therefore energy flows are maximised.

If the dead bodies of phytoplankton are included in the amount of biomass present at the climax stage, then it could be claimed that succession in aquatic ecosystems also tends towards maximum energy storage at climax.

9. Productivity. The gross productivity of the ecosystem increases through succession to climax. This increase is in proportion to the standing crop. The percentage of gross productivity fixed as net productivity does not continue to increase to the climax stage. This is due to several reasons,

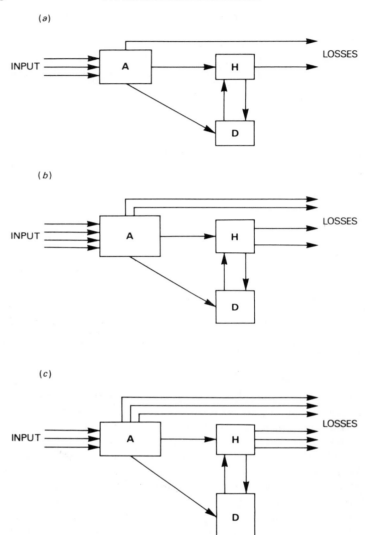

FIG. 16 *Energy flow in ecosystems in various stages of development. (a) successional stages; (b) steady-state of climax; (c) disturbed ecosystem. A: Autotrophs; H: heterotrophs; D: dead organic matter.*

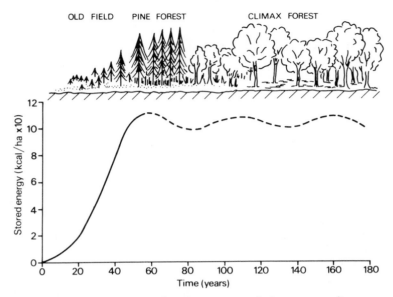

FIG. 17 *Typical pattern of energy accumulation as standing crop in a secondary succession on abandoned pasture, eastern North America. (From* The Ecological Effects of Radiation, *George M. Woodwell. Copyright © 1963 by Scientific American, Inc. All rights reserved.*

(*a*) In the early seral stages the dominant plants tend to be small and shortlived. This type of plant, which includes the annual weeds, has high net productivity. The small bodies of these plants require relatively little energy for maintenance.

(*b*) In the later seral stages the dominant plants tend to be large and longlived such as trees. When fully grown these have to expend a high proportion of their gross productivity in respiration to maintain their bodies. Organisms have their maximum growth rates when young and are characterised by decreasing net productivity when mature. Consequently, the large longlived plants spend a significant period of their lifespan in a relatively unproductive state. This is reflected in the productivity pattern of the ecosystem as a whole.

10. Ecological efficiency.

(*a*) The traditional theory of succession implies that the process leads towards a community which attains the maximum efficiency of energy conversion. Energy is the ultimate limiting resource for ecosystems so it was logical for people to assume that maturity was reached when the best possible use was made of the available energy. However, this idea conflicts with what is known now about patterns of energy-flow and productivity.

(b) We have noted that in a primary succession gross productivity starts from zero and increases as the seral stages develop. Therefore ecological efficiency must also start from zero and increase. However, it cannot increase indefinitely if net productivity declines towards climax. The efficiency of energy conversion decreases in the late seral stages.

(c) The declining ecological efficiency of a mature ecosystem is a function of the productivity patterns of the large, longlived plants of the climax. Many ecologists have wondered why evolution has not resulted in large plants with high productivity throughout their lifespans. The reason seems to be that natural selection operates most strongly through competition between young individuals. A plant has a great advantage if it can grow quickly when it is young and vulnerable. When it is large and established, low net productivity does not matter.

CHANGES IN STRUCTURE AND DIVERSITY

11. Stratification.

(a) The pioneer seres are usually comprised of a discontinuous cover of small plants. In terrestrial ecosystems the spaces between the pioneers become colonised to form a simple layer of short vegetation.

As succession proceeds, the bigger plants of later seral stages form addition layers and assume ecological dominance by shading. Initial colonisers are often excluded but other plants may live below the shrub and tree canopies so that typically, a climax forest formation has a complex vertical structure. For example, there may be four distinct layers of plants; a tree canopy, a shrub layer, a layer of herbaceous plants and a ground layer of mosses.

(b) Exceptions to this complex stratification of the climax community may occur. For instance, if the trees of the canopy cast a dense shade, insufficient light energy may be able to penetrate to support understorey vegetation; this is the case in English beechwoods. The deep shade cast precludes the development of much undergrowth. In the succession to beechwoods, the intermediate seres have a more complex layering structure than the climax formation.

(c) The increase in the complexity of the vertical structure of the ecosystem is accompanied by a spatial segregation of function between the layers. For example in a forest, photosynthesis takes place in the canopy layer and decomposition takes place at ground level or in the soil. Tree trunks transport nutrients back to the canopy.

(d) A similar layering of structure and function occurs during successions in the sea and in lakes. Production occurs in the surface layers whereas most decomposition occurs at the bottom of the water. Nutrients are returned to the surface by mixing currents and wind. Therefore, although different mechanisms of nutrient return are involved, it seems that all ecosystems develop layering of structure and function during succession.

12. Species diversity.

(a) A progressive increase in the number of species present features in the early stages of all successions as more plants and animals colonise the area. Initially, the increase in diversity is rapid. In later seres the rate of increase slows down. The number of different species in the ecosystem may continue to increase until the climax community is formed, but more typically there is a decrease in diversity towards the end of the succession.

(b) The decrease in diversity towards climax occurs because of competition. The dominant plants of late seres are larger, and have larger, more complex life-histories than those of early seres.

This results in the competitive exclusion of a lot of species. Therefore, in successions which result in communities dominated by a few species of plant, the intermediate seres often contain the maximum number of species present at any one time during the succession. For example in the hydrosere at Sweat Mere, Shropshire (see Fig. 18), the number of species of vascular plants (plants with conducting tissue) decreases after the sere dominated by alder.

(c) Species diversity may continue to increase towards climax if the structure of the community and available energy supplies allow continued colonisation. For example, tropical rain-forests have complex structures, are dominated by a wide variety of tree species and have abundant energy supplies. In these forests, more habitats are created and exploited towards climax.

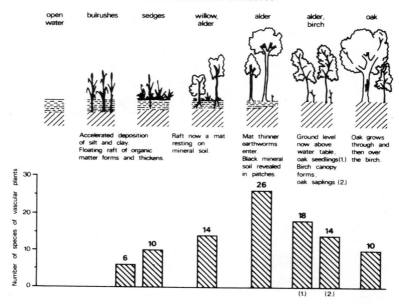

FIG. 18 *Hydrosere at Sweat Mere, Shropshire. (After A.C. Tansley (1939),* **The British Isles and their Vegetation,** *Cambridge University Press).*

13. Trophic structure. Early seral stages have short, linear food chains. These can be disrupted easily because if one link in the chain is eliminated, there is no alternative pathway for energy flow. As the layering of the ecosystem develops and as species diversity increases, the trophic structure becomes more complicated. Larger food chains form and link to create complex food webs.

The more complex trophic structure gives the ecosystem stability. Alternative pathways are available for energy-flow if one link in the food chain is disrupted. Early seral stages are characterised by grazing food chains. The detritus food chains become more important in the later seres when the ecosystem matures.

CHANGES IN NUTRIENT CYCLING

The traditional theory suggests that succession proceeds towards establishing a community that cycles nutrients most efficiently. This is true for most terrestrial successions but is false for most aquatic situations.

14. Nutrient cycling in early seres. In all successions the amounts of nutrients cycling in early seral stages are small. There is little storage in the system. Exchanges of nutrients between the biotic and abiotic parts of the system are rapid because the organisms are short-lived. The role of detritus in nutrient regeneration is unimportant. The organic phases of the cycles are poorly developed, consequently nutrients can move into and out of the system very easily. Nutrient cycling is open.

15. The late stages of terrestrial successions. The increased biomass of late seral stages means that greater amounts of nutrients are stored in the system. The rate of nutrient cycling becomes slower as the system becomes dominated by longer-lived organisms. The amounts of nutrients needed by the late seral stages are greater. The large plants of the climax community have extensive root systems which are very efficient at retrieving nutrients. The role of detritus in nutrient regeneration is important

These characteristics mean that mature systems have the capacity to entrap and hold nutrients for a long time. The organic phases of the cycles are well developed so that there is less import and export of nutrients across the system boundaries. Nutrient cycling becomes more closed or complete. It is relatively efficient and a steady-state is apparent.

16. The late stages of aquatic successions. The amount and rate of nutrient cycling in successions taking place in the sea or large lakes usually declines towards climax so that the sere ends in nutrient-poor maturity. This characteristic develops as a result of the inefficient retrieval of nutrients from bottom sediments. Nutrients released from dead organic matter at the bottom of the water are not returned to the surface productive layer efficiently.

THE CLIMAX CONCEPT

17. The traditional view of the climax community. We have noted that in the traditional theory, ecological succession leads to a final stable community called the climax. This stage has certain attributes. These are, most importantly:

(*a*) It is a stable system in equilibrium with the biotic and physical environment.

(*b*) The species composition is relatively constant.

(*c*) There is no net annual accumulation of organic matter so the standing crop does not fluctuate very much.

(*d*) It is self-perpetuating.

To most ecologists the climax community represents the most efficient use of the habitat, when biomass, energy-flow and nutrient cycling are all at a maximum. There are three main theories about the climax stage.

18. The monoclimax theory.

(*a*) In his original theory about succession in 1916, F.E. Clements stated that the climax community for an area was solely a function of climate. He assumed that, given sufficient time and freedom from interference, a climax vegetation of the same general type would be produced and stabilised within each climatic area. In this way climate determined the boundaries of the climax formations.

This idea became known as the monoclimax theory and was widely accepted by botanists for the first half of this century.

(*b*) Clements and the supporters of this theory could not ignore the fact that many local variations in vegetation communities existed in any one climatic area. These variations were considered to be seral stages however stable they appeared to be. Clements adopted a theoretical long-term view in which local differences in vegetation, for example due to soil conditions, were regarded as exceptions which would be replaced by the regional climax type if sufficient time elapsed.

(*c*) Elaborate special terms were invented to describe these local deviations. For instance, "subclimax" was used to describe a late seral stage which lasted a long time but which would eventually develop to a climax, and "disclimax" was used for a community which replaced a climax after a disturbance such as overgrazing.

So many exceptions to the "monoclimax" type for each climate were recognised and described that numerous workers began to question the fundamental principles of the theory.

19. The polyclimax theory.
Many people thought that the monoclimax theory was too rigid. It did not allow for sufficient explanation of local variations in communities.

In 1939 A. Tansley, a British botanist, proposed an alternative *polyclimax theory*, in which it was possible to have a mosaic of climax types within each climatic area. He realised that climax communities were related to several factors including soil type, drainage and grazing. The polyclimax theory recognises the importance of climate, but maintains that other controls should not be viewed just as temporary phenomena.

The polyclimax theory has the great advantage that all stable communities can be viewed as climaxes. It allows for local variations without the need to categorise them by elaborate terms. The flexibility of this approach was widely appreciated. The polyclimax theory became accepted in favour of the monoclimax theory.

20. The biotic potential theory or climax pattern hypothesis. In the last three decades, ecologists have realised that climax communities are not determined by just one or even several controlling factors. Any particular community is a function of all the interacting factors of the environment such as climate, soil and topography. Therefore as many types of climax can exist as there are combinations of conditions.

This idea was first formulated by R.H. Whittaker in the 1950s. He emphasised that natural communities were adapted to the whole pattern of environmental factors and that climax communities would vary gradually over a region reflecting the gradual changes in factors such as temperature and soil. The climax in any one area reflects the potential for ecosystem development in that location. This idea is known as the *climax pattern hypothesis* or *biotic potential theory*.

This approach is far less abstract than the monoclimax or polyclimax theories. It allows for a more realistic assessment of climax communities.

CHALLENGES TO THE CLIMAX CONCEPT

One of the most recent developments in ecology has been the reassessment of the climax concept as outlined in the previous section. People have questioned the reality of the attributes ascribed to the climax stage. Most workers agree that succession tends towards more stable and complex ecosystems, but they feel that the characteristics of the end product need to be re-examined.

21. The question of stability.

(*a*) The traditional climax concept implies a state of equilibrium with environmental conditions, most importantly with climate. This approach is vulnerable because climates are known to be constantly adjusting and oscillating especially in temperate areas. Therefore, it would be impossible for vegetation to become completely attuned to a particular climate. The only possible exceptions to this are the ecosystems of equatorial regions where only minimal oscillations of climate have occurred. In higher latitudes ecosystems have to adjust to climatic changes associated with repeated glaciations.

(*b*) Even in areas where climates remain fairly constant, the equilibrium of a climax community may be upset by changes in species composition caused by migrations or oscillations in population numbers.

(*c*) In view of these aspects, it is more realistic to consider the climax stage as a community which achieves relative stability rather than equilibrium. Changes will occur in it. They may be directional, following directional changes in climate, for example progressive warming after glaciation.

(*d*) The important difference between the climax stage and previous seres is in the rate of change. In early seres the rates of change are rapid. In the climax stage rates of change become minimal.

22. The question of self-perpetuation.

(*a*) Self-perpetuation is central to the traditional idea of the climax. Very few communities can be seen to be truly self-perpetuating both in structure and species composition. Those which are tend to be ones limited by adverse environmental conditions, for example deserts. These fit into the polyclimax or climax pattern theories.

(*b*) Most climax communities are self-perpetuating in their general structure but not in their species composition. For example, deciduous forest formations may exist in a region for thousands of years but the mixture of dominant trees and associated understorey plants will change, reflecting oscillations in climate.

(*c*) Some climax communities are clearly not self-perpetuating. These may undergo cyclic developments. An example of these occurs in British beechwoods. Beech trees are incapable of regenerating beneath their own canopies. Adult beech trees cast dense shade and produce acidity in the surface layers of the soil. These conditions cannot be tolerated by beech seedlings. Because of this, the dominant beech canopy is eventually replaced by other species such as ash and oak, which formed late seral stages in the succession. Beech seedlings can grow beneath these types of tree, so the climax beech community will be restored. In this way a cycle occurs between the climax and the late seral stages.

IS SUCCESSION AN ORDERLY PROCESS?

If the validity of the traditional view of the climax community is challenged, it follows that the idea of succession as an orderly process leading towards a predictable type of end community must be reconsidered. Two main aspects are important here.

23. Environmental determinism versus the random element. Until recently, most work emphasised the control of the environment in determining which species could inhabit particular areas. This approach is known as environmental determinism. Botanists thought that the communities which would develop in an area could be predicted from the environmental conditions.

It is now widely accepted that succession involves random processes. As early as 1926, H.A. Gleason, an American ecologist, pointed out that all successions and communities could be explained as a result of the random spread and establishment of individual plants. He considered that the apparent orderly changes in species composition, which occurred in a succession, reflected no more than the rate at which available local species could invade the habitat. In his view, the communites were just randomly acquired assortments of plants suited to the neighbourhood.

This idea is undoubtedly correct but obviously the type of plants which can invade an area will alter as environmental conditions change, for example as soil develops.

24. Is the sequence of communities in a succession orderly?

(*a*) If each community is the result of random processes of plant dispersal, doubts are cast on the orderly nature of successions. As we have noted previously in this chapter (**6**), few detailed studies have been conducted on primary successions because of the length of time required for vegetation changes to take place. However, many people have witnessed secondary successions, the invasion of abandoned fields by natural vegetation has been observed frequently. Changes which take place in the general structure of this vegetation as it proceeds through a weed stage to a shrub stage and eventually towards stability at woodland or forest, are well known. The general changes in ecosystem structure can be seen to be orderly and are predictable, although changes in species composition may not be.

(b) It is now realised that the sequence of actual communities in a succession may vary even if changes in general structure are orderly and directional. For example, the sequence of communities in similar hydroseres may not follow a definite pattern.

In 1970, D. Walker studied the order of community succession in ponds. He sampled 66 sites, in which he identified twelve types of plant community. At each community location, Walker examined the plant fossils in the sediments beneath the existing sere to determine what the previous communities had been. He discovered that there was no orderly or predictable sequence of communities in the hydroseres. Community succession appeared to proceed in a random fashion. There may be several different paths by which succession may proceed at any one time in a given area.

(c) In view of these results, many botanists are convinced that succession should be considered as a phenomenon involving the properties and characteristics of individual plants rather than as a sequence of district communities.

SUCCESSION VIEWED AS THE REPLACEMENT OF OPPORTUNIST WITH EQUILIBRIUM SPECIES

Once ecological succession is seen to be the result of the dispersal and establishment of individual plants rather than as the development of discrete communities, it becomes possible to view it in the wider context of the strategies for survival adopted by individual species.

These strategies can be divided into two main groups. Firstly, the opportunist group which is adapted to exploit the open environment of the developing ecosystem, and secondly the equilibrium group, which is adapted to exploit the conditions of mature ecosystems. An examination of the changes which take place at the species level provides a better understanding of the causes and inherent nature of succession.

25. The opportunist strategy.

(*a*) *The pioneer plants are opportunists*. They are adapted to colonise the bare ground of open habitats. They produce a large number of seeds which are dispersed easily. In order to do this, they must be highly productive and devote most of their energy to reproduction.

(*b*) *Opportunist species are small*. This is partly because most of their net productivity is devoted to seed production but also because there is no need for them to grow large bodies. Competition between individual plants is at a minimum on bare ground sites. Tallness is of no great advantage in this habitat.

(*c*) *Opportunist species are shortlived*. Typically, they are annual plants (i.e. have lifecycles that are completed within one growing season). This enables them to put a maximum amount of energy into reproduction rather than divert some to produce a plant body, such as a bulb or a shrub, which could withstand the adverse conditions of winter.

(*d*) *Opportunist species are generalists*. They can tolerate a wide variety of environmental conditions, especially of soil type, temperature and moisture. However, they usually require open habitats because they cannot tolerate shading.

26. The equilibrium strategy.

(*a*) The species of the late seral stages and climax communities are equilibrium species. They are adapted to live in relatively stable and predictable environments.

(*b*) Equilibrium species can compete effectively against other species in the habitat. In order to do this they must achieve ecological dominance. They grow tall and are longlived (typically they are perennials, i.e. live for many years). Equilibrium species devote a lot of their net productivity to building and maintaining a large plant body.

(*c*) Equilibrium species usually have poor powers of dispersal. They produce few seeds which tend to be relatively large; consequently they extend their distribution ranges slowly.

(*d*) Equilibrium species are specialists. They are specialised for exploiting the conditions prevailing in a particular habitat. This means that they can compete well within certain environments but that they cannot tolerate the wide range of conditions which could be encountered on a bare site.

During succession the opportunist species are gradually

replaced by the more persistent equilibrium species. These attain
ecological dominance and shade out the pioneer plants.

THE IMPORTANCE OF SUCCESSION FOR HUMAN FOOD PRODUCTION

The succession concept is a direct relevance to many of man's
activities. Most importantly, in agriculture our goal of maximum
production leads to basic ecological conflicts.

27. Types of plants cultivated. Most of our crops are opportunist
species of early and middle seral stages. These are the plants
which grow quickest on bare ground. They do not have large,
inedible structures such as vast, woody root systems but put a lot
of their productivity into reproductive structures like cereal
grains. Thus, they yield a concentrated food source for human
consumption.

Some crops could be described as post-pioneer species. For
example, some such as sugar-beet and potatoes, do have
underground storage organs but these are still species of early
successional stages. The majority of the trees used for fruit or
timber are characteristic of middle seres. They would not survive
in a climax forest. All these types of plant have high net
productivity and relatively short life-cycles.

28. Implications for stability. Since our agricultural ecosystems
are like early seral stages, they are instable. The non-equilibrium
communities have to be maintained artificially. Activities such as
weeding and spraying with insecticide to protect the crop from
pests and diseases, require the expenditure of energy subsidies as
fuel.

Farming the pioneer stages invokes the practical cost of
nutrient loss. Nutrient cycles are open in early seral stages so there
will be appreciable losses from the system. These must be
returned in the form of fertilisers.

As farmers take over more land, succession is driven back to the
early seres in more areas. The removal of complex mature
ecosystems not only causes mass extinctions of species which
require this type of habitat, but also disrupts biogeochemical
cycles on a large scale. Elimination of mature, relatively stable
communities can cause soil erosion and upset nutrient cycling.
The significance of these aspects is discussed further in XIV.

PROGRESS TEST 4

1. What are the main causes of change in ecosystems? **(1)**

2. Explain the difference between primary and secondary succession? **(6)**

3. What is the traditional theory of ecological succession? **(6)**

4. How do energy flow patterns change during succession? **(8)**

5. Why does net productivity decrease as ecosystems approach stable state or climax? **(9)**

6. Describe the development of structural and functional stratification in ecosystems during succession. **(11)**

7. Why does species diversity usually decrease as ecosystems approach stable state or climax? **(12)**

8. Compare the pattern of nutrient cycling in early seres with that typical of late seres. **(14–16)**

9. What are the differences between the monoclimax and polyclimax theories? **(18,19)**

10. Why has the concept of the climax been challenged? **(21,22)**

11. Describe the opportunist strategy of plant survival. **(25)**

12. What is the importance of succession for human food production? **(27,28)**

CHAPTER V

Environmental Factors 1:
Introduction, Light and Temperature

INTRODUCTION

Factors which have an effect on the life of an organism during some stage of its development are called *environmental factors.* The plants and animals forming the biotic part of the ecosystem will be those species which can tolerate the prevailing environmental conditions. No organism exists in isolation. Each must be considered in the context of its environment as this determines the conditions for life.

1. Types of environmental factor. Environmental factors can be divided into four groups:

(*a*) *Climatic factors.* These include the main climatic parameters of light, temperature, water availability and wind.

(*b*) *Edaphic factors.* These are the characteristics of the soil such as nutrient status, acidity and moisture content.

(*c*) *Topographic factors.* This group includes the influence of terrain features, for example slope angle, slope aspect and altitude.

(*d*) *Biotic factors.* These are all the interactions of living organisms such as competition, grazing and shading.

2. Relationships between the environmental factors. It is obvious that in studying ecosystems, it is essential to analyse how the environmental factors are operating. In practice, the four groups are so interrelated that it is extremely difficult to isolate the influence of individual environmental factors. For example, both topography and climate will influence soil development. Similarly both climate and soil will exert strong influences on the pattern of biotic controls which operate in an ecosystem by determining which species can inhabit the area.

The environmental factors include both biotic and abiotic controls. However, the fundamental characteristics of any ecosystem will be governed by its abiotic components. The effects of these variables may be modified by the plants and animals of the system, for instance by trees creating shelter from strong winds, but the extent of this modification is limited.

The abiotic factors are basic controls on the ecosystem as a whole. The biotic controls are, nevertheless, still very important because they influence the distribution and function of individual species.

All the environmental factors vary through time and space. Living organisms exhibit definite responses to this variation so distinct relationships exist between the environmental factors prevailing in a specific ecosystem and the communities formed there. These relationships are often very intricate because complex interactions take place between the organisms and their environments and also amongst the organisms themselves.

In addition, historical factors are often involved. The ecosystems may be adjusting to comparatively recent changes in climate. Alternatively, some aspects of their species content and function may be a legacy of ancient events in evolution. The influence of these past events is discussed in VIII and X.

LAW OF THE MINIMUM

3. Liebig's law of the minimum.

(a) In 1840 Justus Liebig, a German chemist, conducted some pioneer studies on the effects of various factors on crop growth. He found that the yield of a crop was often limited by nutrients needed in scant supply rather than by those required in large amounts like carbon and water. For example, he discovered that phosphorus deficiency was frequently the factor which limited growth. This led him to realise that just one overriding factor may be limiting the productivity of a plant.

(b) Following the scientific fashion of his time, he presented his conclusions in the form of a law, known as the "law of the minimum". This states that the growth of a plant depends on the amount of foodstuff presented to it in limiting quantity. So, for instance in agriculture, crop growth often depends on the amount of phosphorus present as this is frequently the substance present in limiting quantities.

4. Modifications of Liebig's law. The law of the minimum is best applied only to the chemical materials necessary for growth and reproduction. Liebig did not intend it to include other

environmental factors, although many susbequent workers have expanded his statement to encompass things like temperature and light.

Work since the time of Liebig has shown that two ancillary clauses must be added to the concept to make it useful in practice.

(a) Firstly, it is applicable only under steady-state conditions. Unless the inputs and outputs of energy and materials of the ecosystem are in approximate balance, the amounts of the various substances required will be constantly changing and the law cannot be applied.

(b) Secondly, the law has to take account of factor interaction. A high concentration or availability of one substance may modify the rate of utilisation of the substance in minimum supply. Occasionally organisms are able to use a substitute chemical, closely related to the one that is deficient in the habitat. For example, if calcium is lacking but strontium is abundant, some molluscs are able to use strontium for shell construction instead of calcium.

LIMITS AND TOLERANCES

5. Shelford's laws of tolerance.

(a) One of the most significant developments in the study of environmental factors occurred in 1913 when Victor Shelford proposed his laws of tolerance. In these laws he outlined the importance of tolerance in explaining the distributions of species.

The laws state that for each environmental factor a species has a minimum and a maximum condition which it can endure. Between these extremes there is a range of tolerance which include an optimum condition.

(b) The ranges of tolerance can be represented as bell shaped curves (see Fig. 19) The range of tolerance can vary both between species for each environmental factor and also within one species for different factors. For example, one species (species A in Fig. 19) may have a much wider range of tolerance to temperature than another species (species B in Fig. 19). Similarly, an individual species may have a wide range of tolerance to temperature but a narrow range of tolerance to soil conditions.

(c) A series of terms have come into general use in ecology to describe the ranges of tolerance. These terms use the prefix steno

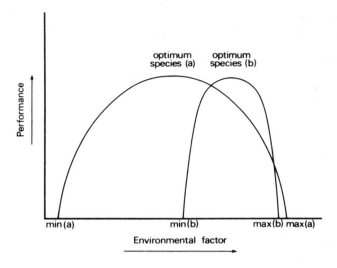

FIG. 19   *Examples of bell-shaped tolerance curves.*

(from the Greek *stenos*), if the range of tolerance is narrow, and the prefix eury (from the Greek *eurys*), if the range of tolerance is wide. The most frequently used of these terms are given in Table IV.

TABLE IV. TERMS USED TO DESCRIBE VARIOUS KINDS OF TOLERANCE

Narrow Range	Wide Range	Factor referred to
Stenothermal	eurythermal	temperature
Stenohydric	euryhydric	water
Stenohaline	euryhaline	salinity
Stenophagic	euryphagic	food
Stenoedaphic	euryedaphic	soil
Stenoecious	euryecious	habitat selection

(*d*) Shelford's laws state that those species with wide ranges of tolerance for all factors will be the ones which are most widely distributed.

(e) In addition Shelford stated that environmental factors are most limiting during the reproductive stages of lifecycles. Seeds, eggs, embryos and also the reproductive adult animals all have narrower ranges of tolerance than do the mature or non-reproductive stages.

6. The use of toleration ecology. The work of Shelford inspired many studies in toleration ecology. Numerous stress tests were conducted in laboratories to determine the tolerance ranges of individual species for the different environmental factors. Although this work was useful in some applied aspects, such as determining the tolerance of fish to water pollution, it did little to help explain natural distributions. Shelford himself had said in his laws that the reaction of an organism to one environmental factor may be related to the condition of another factor; for example, when soil nitrate is limiting, the resistance of grass to drought is reduced. Therefore, the artificial study of individual factors provides only part of the story.

7. Optimum locations. It is surprising that Shelford also realised that plants and animals rarely live at their optimum locations. Organisms are often excluded from their optimum habitats by competition from other species, so they exist where they can compete most effectively. For instance, many desert plants grow better in moist climates but they are confined to deserts because these are the habitats in which they have the greatest ecological advantage.

Another reason why species inhabit suboptimal locations is because environments are constantly changing. Constant shifts in location would be necessary if species were to always inhabit their optimum environments.

8. Combined concept of limiting factors. Although Shelford's laws are basically correct most ecologists now consider them to be too rigid. It is more useful to join the idea of the minimum with the ideas of toleration to give a general combined concept of limiting factors. This recognises that the presence and success of an organism depends on a complex of conditions.

Organisms are controlled in nature not only by the supply of materials for which there is a minimum requirement but also by the other environmental factors which are critical. Any factor which approaches or exceeds the range of tolerance may be a limiting factor in the distribution of the species.

It is often difficult to define which factors are limiting a distribution in nature because of the problems involved in isolating the influences of the individual components of the habitat. The chief value of the combined concept of limiting factors is that it gives a framework for the study of complex relationships.

9. An evolutionary view of tolerance. Many ecologists now consider that too much attention has been focused on the study of the tolerance ranges and the limiting factors themselves. Instead, they feel that scientists should examine the ways in which plants and animals have evolved to exploit particular habitats, and hence have had to develop suitable ranges of tolerance to the environmental factors in order to survive.

The work in toleration ecology inspired by Liebig and Shelford does not answer the basic ecological question of why species have adapted to their particular sets of limits. A more advantageous ecological viewpoint is to consider how species have evolved to get a living and to regard tolerance ranges as byproducts of the requirements for a chosen way of life. This approach, which emphasises the importance of evolution, has led to a better appreciation of the relationships between individual species and habitats.

The climatic, topographic and edaphic factors are examined in the rest of this chapter and in VI. The complex biotic factors are discussed in VII, VIII and IX in the context of demography and evolution.

LIGHT

10. The importance of light as an environmental factor. Light is a vital factor in the environment because it is the ultimate source of energy in all ecosystems. We have noted in the previous chapters that the structure and function of ecosystems is largely dictated by the amount of incoming solar radiation. However, light can be a limiting factor in excess. Exposure to too much light can damage and even destroy living tissue. Light varies in three important aspects:

(*a*) quality or wavelength composition;
(*b*) intensity or energy content;
(*c*) duration, particularly day-length, i.e., the number of hours of light received each day.

Variation in these three parameters governs all sorts of physiological and morphological processes in both plants and animals. Although the influence of light is often linked with that of other factors, such as temperature and water supply, its specific effects are frequently the most important controls in the environment.

11. Variations in the quality of light.

(a) Solar radiation consists of electromagnetic waves of a wide range in length. Not all of them can penetrate the upper atmosphere to reach the surface of the earth. Those which do consist of waves between 0.3 and 10 microns (μ) in length ($1\mu = 1/$1000th of a mm). To the human eye, visible light (i.e. white light) lies in the range between 0.39 and 7.60 μ. Solar wavelengths below 0.39 μ are short-wave radiation, known as ultraviolet light. Those above 7.6 μ wavelength are long-wave radiation, known as infrared light.

(b) In general the quality of light does not vary significantly from one part of the biosphere to another, so it does not usually constitute an important ecological factor. However, both animals and plants are known to respond to the different wavelengths of light.

12. The importance of light quality.

(a) Most plants are adapted to exploit light of wavelengths between 0.39 and 7.60 μ. Ultraviolet and infrared radiation are not used in photosynthesis. As chlorophyll is green, it absorbs the red and blue light, so these wavelengths are the most important part of the visible spectrum for photosynthesis.

(b) In terrestrial ecosystems the quality of light does not vary sufficiently to exert an important influence on the rate of photosynthesis except where vegetation canopies intercept a lot of sunlight. The light penetrating through to the understorey plants will not only be less intense than that impinging on the canopy, but it will also be deficient in the red and blue wavelength. Plants living in these habitats must be adapted to a low energy environment.

(c) In aquatic ecosystems red and blue light is filtered out by the phytoplankton living near the surface. The resultant greenish light, which penetrates to lower layers, is poorly absorbed by chlorophyll. The red algae (which include many of the seaweeds) have additional brown-red pigments (the phycoerythrins) which can absorb green light for use in photosynthesis. This enables these algae to live at greater depths.

(d) The effect of ultraviolet light on plants is still obscure. It is known to be harmful to bacteria and it is believed to exert a retarding effect on vegetation development. Much short-wave radiation is absorbed in the upper atmosphere so only a small percentage reaches the earth's surface. For this reason the influence of ultraviolet light is greatest at high altitudes.

The rosette leaf-form, characteristic of many mountain plants, is thought to be a result of ultraviolet light preventing the elongation of the stem. The influence of ultraviolet radiation probably prevents many species of plant from migrating across mountain barriers. In this way it acts as agent in determining distribution patterns.

(e) The role of light quality in the distribution and function of animals is uncertain. Colour vision in animals has an interesting spasmodic distribution within the different taxonomic groups. Apparently, it is well developed in some species of arthropods, fish, birds and mammals but not in others of the same groups. For example, among the mammals, colour vision is well developed only in the primates.

13. Variations in the intensity of light.

(a) The intensity, or energy content, is the most important aspect of light as an environmental factor because it is the ultimate driving force of the ecosystem. Light intensity varies greatly both spatially and temporally.

(b) As solar radiation passes through the earth's atmosphere some light wil be absorbed, reflected or scattered by gases and particles. Light intensities are greatest in the tropics especially in the arid zones where very little light is reflected by cloud cover. In the low latitudes the sun's rays penetrate the atmosphere at a high angle to the earth's surface so passing through a minimum depth of atmosphere.

Light intensities decrease progressively with increasing latitude. At high latitudes the sun's rays are at a lower angle to the surface of the earth, and so have to pass through a greater depth of atmosphere, suffering more depletion of radiation en route. In addition more light is reflected and scattered at high latitudes by the frequent cloud covers and polluted atmospheres.

(c) The basic global variation in light intensity is complicated by seasonal variations. Light intensities in the high latitudes vary greatly between summer and winter.

The large-scale variations in light intensity can be modified further by topography. The angle and orientation of slopes have significant influences on the amount of light energy received by a ecosystem. For example in high latitudes in the northern hemisphere, a steep south-facing slope would receive far more insolation than a steep north-facing slope and would probably develop a different ecosystem structure.

14. The importance of light intensity variations.

(*a*) Light intensities vary within ecosystems. Vegetation canopies intercept and absorb some light so determining the amount of energy available for use by the understorey plants. The vertical stratification of an ecosystem is thus a product both of the total light energy available and of the community itself.

(*b*) In aquatic ecosystems the intensity of light decreases rapidly with increasing depth. Water reflects and absorbs light very efficiently. In clear, still water only about 50 per cent of the light impinging on the surface reaches to below 15 m depth. If the water is moving or is turbid, far less will penetrate to depth. This imposes a severe restraint on photosynthesis.

(*c*) Excessively high light intensities can act as a limiting environmental factor. Very intense light can damage enzymes by causing photo-oxidation. This impairs the metabolism of organisms especially their ability to synthesise proteins.

15. Compensation points. In order for a plant to have net productivity it must receive enough light energy to manufacture sufficient carbohydrates to compensate for those used up during respiration. If all other variables which influence the rates of photosynthesis and respiration are assumed to be constant, the balance between the two processes can be shown in relation to light intensity (*see* Fig. 20).

As light intensity increases, photosynthesis will increase until a plateau is reached at the maximum amount possible for the plant. The amount of carbohydrates used up in respiration by the plant is marked on Fig. 20(*a*) and (*b*) as a broken line. The point at which the rate of photosynthesis (carbohydrate formation) crosses the line for the rate of respiration (carbohydrate usage) is known as the compensation point. This is the point at which light intensity is sufficient for net productivity to occur, and this is the minimum light intensity essential for growth. The compensation point varies for different types of plant.

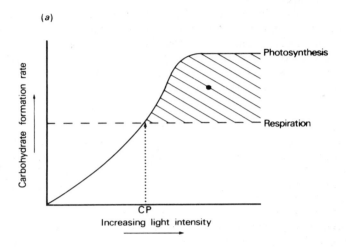

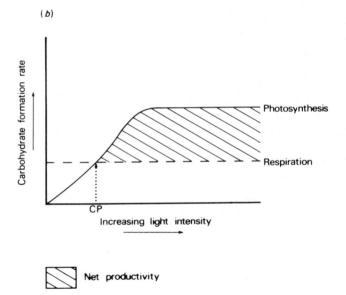

FIG. 20 *Diagrams showing the compensation points of (a) heliophytes (sun-plants) and (b) sciophytes (shade-plants).* **CP**: *Compensation point.*

16. Sun-plants and shade-plants. Plants adapted for life in high light intensities have high compensation points. This type of plant is known as a sun-plant or *heliophyte*. Its body chemistry is very active so it uses up a lot of carbohydrates in respiration (*see* Fig. 20 (*a*)). Plants adapted for life in low light intensities have low compensation points. This type of plant is known as a shade-plant or *sciophyte*. Its metabolism tends to be slow so it has a low respiration rate (*see* Fig. 20 (*b*)).

Some species of plant have the characteristics of sciophytes when they are young, then develop the characteristics of heliophytes as they mature. This is a particular advantage to the tree species whose seedlings must survive in the shade of the forest floor. Once the young tree has matured it can exploit the high light intensity impinging on the canopy layer.

17. The importance of light duration. The relative length of days and nights (i.e., the duration of daylight in each 24 hours) effects the function of a wide range of plants and animals. The response of an organism to varying day lengths is known as *photoperiodism*. In plants the responses include the initiation of flowering, leaf-fall and dormancy, while in animals they include the initiation of migrations, hibernation, nesting and changes in coat colour.

18. Variations in day length. In the equatorial regions day lengths (or photoperiods) are of a constant 12 hours duration throughout the year. In the temperate zones day lengths are greater than 12 hours in the summer but shorter than 12 hours in the winter. The greatest annual differences in day length occur in the very high latitudes. These experience continuous daylight in the summer months and continuous darkness throughout the winter.

19. Photoperiodism in plants. In plants the reaction to changing day length is often linked to changes in temperature. However, in many instances the relative duration of light and dark periods can be shown to be the critical stimulus inducing the response.

Flowering plants can be divided into three broad groups on the basis of their photoperiodic reactions.

(*a*) *Long-day plants.* These need day lengths greater than 12 hours duration for flowering to occur. They include many temperate crops such as wheat, barley and spinach.

(*b*) *Short-day plants.* These require day lengths of less than 12 hours duration for flowering to occur. They include tobacco and chrysanthemums.

(c) *Day-neutral plants.* These do not have any particular requirements for day length periods to initiate flowering. This group includes tomato and dandelion.

The reactions of the long and short day plants limit their distribution to the latitudes where suitable photoperiod conditions occur. If some species of plant are deprived of their optimum day lengths they show increased vegetative growth. For example, short-day onions will produce larger bulbs when grown under conditions of long-days. This has an economic value and has been utilised by horticulturists.

20. Photoperiodism in birds. In the vertebrate groups the most highly evolved photoperiod reactions are found in the birds. Responses to changing photoperiods are known to cause the start of migrations in many species of bird. For this reason the migrations of birds from the temperate zones often start within the same few days in consecutive years for each species irrespective of weather conditions.

Similarly the breeding cycles of many types of bird are initiated by changes in day length. The sex organs (gonads) of birds are at their minimum size in the winter months when the birds are in a non-reproductive state. The gonads start to expand in the early spring. Their rate of growth accelerates as day-lengths increase until the maximum size is reached at the time of year when the birds reproduce. After reproduction the gonads decrease in size again. In many birds this annual redevelopment of the gonads depends rigidly on the return of longer days in springtime.

Within the breeding cycle egg production can be manipulated by altering the photoperiod. It has been known for over 200 years that egg production in hens could be increased by subjecting the birds to artificial light at night. In this way the birds experience two photoperiods in every 24 hours and so lay two eggs each day instead of one. This is now an established practice in poultry husbandry.

21. Photoperiodism in mammals. This group has received far less attention than the birds, partly because their photoperiod responses are generally less rigid.

(a) *Long-day responses.* Many wild mammals, especially those of higher latitudes, breed in response to the increasingly long photoperiods of spring. This response has been demonstrated experimentally in ferrets, hares, hedgehogs and white-footed mice. It seems to be a synchronising factor in the seasonal regulation of breeding cycles.

(b) *Short-day responses.* Many breeds of sheep, goats and deer develop their reproductive activity in response to the short days at the end of summer and early autumn. Mating takes place at this time so that the young are born in spring when conditions are most favourable. Breeding activity stops when the photoperiods start to increase again.

22. Photoperiodism in fishes. A wide range of external stimuli coordinate the seasonal breeding and migratory activities of fish but heat and light are known to be the most important environmental factors in these activities. Fish living in the light zones near the surface respond to photoperiods in a variety of ways. For example, it can be shown experimentally that light affects the reproductive organs of fish. Trout exhibit an increased reproductive capacity if subjected to artificially long photoperiods. This phenomenon has been used in commercial trout fisheries.

Photoperiods also influence fish migrations by acting on the organisms' hormonal systems. For instance, in the three-spined stickleback hormonal changes induced by photoperiods influence the salinity preferences of the fish. These cause it to migrate from the sea to fresh water in the spring and then back to the sea again in the autumn.

23. Photoperiodism in insects. Although this group of animals is generally slower to respond to changes in environmental factors it still exhibits photoperiod reactions. An example of this can be seen in the life cycle of the colorado beetle which feeds on potato plants. The beetle hibernates in the soil in response to the short days of autumn, thus avoiding the adverse conditions of winter when its food supply is not available.

In many insect species metamorphosis is influenced by day length. In particular the time of day of emergence from the pupae is affected by the photoperiod.

TEMPERATURE

24. The importance of temperature as an environmental factor. Temperature acts as an environmental factor both directly and indirectly. It has a direct effect on nearly every function of plants and cold blooded animals by controlling the rate of their body chemistry. Temperature acts indirectly by influencing the condition of other factors, especially water supply. Temperature determines the rate of evaporation and hence not only the effectiveness of rainfall but also the rate of water loss from organisms.

It is often difficult to isolate the influence of temperature as an environmental factor. For example, the energy of light may be converted to heat energy when the light rays are absorbed by a substance. In addition, temperature often acts in conjunction with both light and water to control the function of organisms.

It is relatively easy to measure temperatures in the environment but it is difficult to decide which temperatures are most significant. In many cases it is not obvious whether the maximum, minimum or average temperatures are most important.

25. Variations in temperature.

(*a*) Few places in the world experience temperatures which are continuously too hot or too cold for life to exist but temperatures vary a great deal both spatially and temporally. Large-scale variations occur with latitude. Superimposed on these differences, there are local variations due to both topography and distance from the oceans. For example, in the northern hemisphere, south-facing slopes are warmer than north-facing ones.

(*b*) In addition, there are variations in temperature within eco-systems, particularly within forests and within aquatic environments. Significant differences in temperature can occur between the canopy layer of a forest and the forest floor, and between the surface layer of a lake and the water at depth.

Variations in temperature through time occur both seasonally and diurnally (i.e., daytime to nighttime differences). All these variations influence the distribution and function of organisms.

TEMPERATURE AND PLANTS

26. Range of temperature tolerance in plants. Most life in the biosphere functions within a temperature range of between 0°C and 50°C. Within this range, individual species have minimum, maximum and optimum temperature requirements for their metabolic activities. These particular temperature requirements are known as *cardinal temperatures.*

The temperature of a plant body is usually approximately the same as that of the environment because there is a constant transfer of heat between the plant body and the air.

The ranges of tolerance to temperature in plants vary greatly. For example, tropical crops such as melons cannot tolerate temperatures below about 15°C – 18°C, whereas temperate cereals cannot tolerate temperatures below minus 2°C to minus 5°C. In contrast, evergreen conifers can tolerate temperatures as low as minus 30°C. Aquatic plants generally have narrower tolerance ranges than land plants.

With all species of plant, the exact tolerance ranges vary with factors such as the age of the plant, its water balance and also the season of the year.

27. Damage to plants caused by excess heat. The maximum temperatures tolerated by plants are often more critical than the lower limits. Plants are usually cooled by the loss of water from their bodies. Therefore, heat damage occurs only if insufficient water is available for cooling. In most cases damage induced by high temperatures is associated with damage caused by water deficiency, e.g. wilting. In these circumstances enzymes become inactive and the plant's metabolism is slowed down.

Plants which inhabit hot climates frequently have morphological (i.e. structural) adaptations to enable them to survive. These include thick corky bark which acts as an insulating layer, small leaves to reduce water loss and thick cuticles (the outside covering of plants) to reflect light.

28. Damage to plants caused by excess cold.

(*a*) Most plants cease growth at temperatures below 6°C. Further decreases in temperature below this level may cause serious injury. Proteins may be precipitated out of solution in the cell sap rendering enzymes inactive. If temperatures reach freezing point, ice forms between the cells of the plant drawing water out of the cells themselves causing dehydration. If rapid freezing occurs ice crystals can form within the cells. The crystals are large compared to the size of the cell. As they form, they puncture the cell's membranes which leads to the death of the cell. This produces the brown areas on plants, characteristic of frost damage.

(*b*) Low temperatures may also act indirectly to inhibit the function of plants. Roots are less permeable when cold and so are unable to take in water. This can lead to a *physiological drought* which occurs when water is present but cannot be taken into the plant because the temperature is too low. It is particularly important in tundra environments (*see* XI).

(*c*) Plants which inhabit cold climates frequently have morphological adaptations to enable them to survive. These include being of dwarf stature or creeping habit so that they are less exposed, or having cushion or mat forms so that one part of the plant can give protection to other parts.

29. Temperature and plant lifecycles. Many plants have evolved lifecycles which enable them to endure cold seasons. These adaptations can be of three basic types.

(*a*) *Annual plants.* These have short lifecycles which are completed in the warm season. The plants overwinter as seeds which are tolerant of cold.

(*b*) *Herbaceous perennials.* These plants have a resistant storage organ in the soil such as a bulb, a corm or a rhizome. The plant survies through the winter as the storage organ and produces a new set of shoots every year.

(*c*) *Woody perennials.* These are the trees and shrubs which have woody structures persisting for many years. They avoid cold damage in one of two ways.

(i) *Deciduous habit.* Leaves are shed at the beginning of the cold season. The buds remaining on the plant are cold resistant.

(ii) *Evergreen habit.* Leaves are retained throughout the year but their resistance to cold is increased by making the cell sap more concentrated. This lowers its freezing point and so reduces the plant's susceptibility to frost damage.

All of these adaptations to survive a cold season involve a period of dormancy.

30. Temperature and dormancy in plants. Dormancy occurs not only in plants of cold environments but also in those of warm climates. Tropical trees often have dormant phases unrelated to temperature. It has been suggested that this phenomenon allowed the ancestors of temperate trees to migrate from the tropics to the temperate zones.

In most cases dormancy is induced in temperate trees in response to photoperiods. However, the dormant phase in these and most other types of plant often needs to be broken by exposure to cold. This is known as *vernalisation*. If insufficient cold is applied to break the dormancy the plant will not grow again. Most trees and shrubs in Britain require between 200 and

300 hours below 9 °C to break their dormancy.

Vernalisation has been utilised in horticulture to speed up lifecycles for breeding purposes. Biennials, i.e., plants with lifecycles extending over 2 years) such as beet and celery, produce leaves and tubers in their first growing season and flowers in their second. If subjected to artificial cold treatment the lifecycle can be completed in 1 year.

31. Temperature and productivity. The rates of respiration and photosynthesis in a plant must be in correct balance for net productivity to occur. For any one species, the optimum temperature for respiration is higher than that for photosynthesis (*see* Fig. 21). Above a certain temperature respiration will outstrip photosynthesis and the plant will starve. This has an important check on the migration of plants from cold to warm areas.

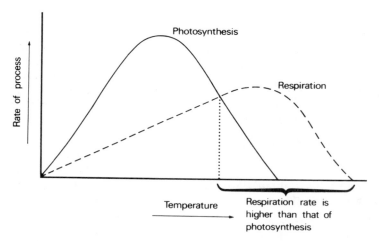

FIG. 21 *The balance between photosynthesis and respiration in relation to changes in temperature.*

32. Thermoperiodism. This is the response of plants to rhythmic fluctuations in temperature. It can occur either on a seasonal or a diurnal basis. Plants which usually grow in habitats that have temperature fluctuations tend to have depressed growth if kept in constant temperatures. Most species grow best if the temperature of their environment is varied. For example, tomatoes have optimum growth rates if they are exposed to day temperatures of 20°C and night temperatures of 10°C. This fluctuation produces the optimum balance between respiration and photosynthesis. Some species require night temperatures to be below a certain minimum for flowering to occur. For instance, the common daisy cannot reproduce unless days are cool and night temperatures are between 8°C and 13°C. In some species regular fluctuations in temperature are required to initiate germination. An example of this is found in *Poa pratensis* (smooth meadow grass) which needs marked fluctuations of day and night temperatures for the seeds to germinate.

Thermoperiodism limits the latitudinal and altitudinal distribution of many plants.

GROWING SEASONS

The growing season is the period when all the environmental conditions required for growth are satisfied. Temperature is regarded as one of the most critical factors in determining the length of the growing season, especially in mid and high latitudes. Mean daily or mean monthly air temperatures are used often in defining the growing season for a particular area. Three methods are commonly used for doing this.

33. Minimum temperatures for growth. The growing season is defined as the period when temperatures are above a certain threshold required for growth. The thresholds used vary from 0°C to 10°C, but frequently 6°C is taken as the limit since this is the minimum temperature necessary for the growth of most temperate crops. In the United States the growing season is often delimited in terms of "frost-free days", that is, the number of consecutive days during which temperatures are continuously above 0°C.

34. Accumulated temperatures. The use of threshold temperatures provides a measure of the length of the growing season but it does not indicate the amount of heat received. This can be done by expressing the quality of heat or the efficiency of the growing season by accumulated temperatures. These are calculated by summing the mean daily temperatures ("day degrees") above a selected minimum temperature. For example, Aberdeen and Chicago both have an average of 7 months with temperatures above 6 °C. During this season, Aberdeen has 1095 day degrees but Chicago has 2328 day degrees.

35. Photothermal units. In high latitudes the shorter and cooler growing seasons are compensated for partly by the longer day lengths. In these situations it is useful to include day light as a factor in defining the growing season. This is done by multiplying the day degrees by the average hours of daylight, giving a measure of the photothermal units of the season.

These methods of delimiting the growing season are not entirely satisfactory. They do not take account of the fact that different plants have different temperature requirements, or that actual air temperatures are modified by soil, topography and vegetation. In addition, the season may be limited by other environmental factors especially water (*see* VI).

TEMPERATURE AND ANIMALS

36. Nomenclature. It is common practice to divide the animal kingdom into "warmblooded" and "coldblooded" species. This subjective division is generally unsatisfactory. A more useful approach is to divide the kingdom into *homeothermic* forms (meaning those with constant temperatures, from the Greek *homoios* = like) which regulate their body temperatures, and *poikilothermic* forms (meaning those with varying temperatures, from the Greek *poikilos* = various) which do not regulate their body temperatures.

37. Poikilothermic animals. These are the invertebrates, fish, amphibia and reptiles. Since they have no mechanism for internal temperature control their body temperatures generally approximate to that of the environment. This means that their activities are governed largely by the state of the external temperature. If temperatures are within a suitable range the poikilothermic animals can function. If it is too cold for activity the animals may survive in an inactive state but survival is limited in excessively hot conditions.

The poikilothermic animals do have some control on their body temperatures by their behaviour and location. For example desert snakes may seek shelter beneath rocks if they become too hot or expose themselves to the sun if they become too cold.

38. Homeothermic animals. These are the birds and mammals.

They have evolved complex means of maintaining their body temperatures within narrow limits (for example, most mammals have body temperatures between 36 °C and 38 °C), so they can remain active despite variations in environmental temperatures. This evolutionary advance gives the homeothermic animals a great advantage over the poikilothermic forms.

The homeothermic animals regulate their body temperatures in a variety of ways both by their morphology and by their metabolism. For instance heat can be retained by insulating layers of fur, feathers or fat or by decreasing the blood flow to the skin surface. Conversely heat may be lost by panting, sweating or increasing the blood flow to the surface of the skin.

39. Damage to animals caused by excess heat. Most animals cannot tolerate environmental temperatures above 45 °C, although much lower temperatures may be lethal for species inhabiting cool environments. For instance, some arctic forms are killed by temperatures as low as 10 °C.

The actual cause of heat death is obscure. It is known that at high temperatures proteins coagulate rendering enzymes inactive. This halts the body chemistry. However, in land animals heat death is usually associated with lack of water; in aquatic poikilothermic animals heat death can be caused by suffocation. As the temperature increases, the metabolic rate of the animal increases so that it needs more oxygen for respiration. If sufficient oxygen cannot be obtained from the water the animal will die.

Many animals have evolved methods of avoiding damage by excess heat, including ways of retaining body fluids by, for example, producing concentrated urine, and ways of avoiding the severest temperatures such as by being nocturnal or by having a dormant period at the hottest time of year. This type of dormancy is known as *aestivation*.

40. Damage to animals caused by excess cold. As the temperature of an animal's body decreases, its metabolic rate decreases and the animal becomes lethargic and slow. Many animals tolerate temperatures below 0 °C, but if the body tissues of vertebrates freeze extensive damage is caused by the ice crystals in the cells. This damage or "frostbite" is irreversible and can be lethal. Strangely, the bodies of some invertebrates can be frozen without apparent damage, for example, mosquito larvae are frozen in Alaskan ponds without harm.

Evolution has adapted many animals for life in habitats with very cold seasons. Some species have efficient methods of heat retention such as thick fur and fat layers, whilst others avoid the cold by becoming dormant. True hibernation occurs only in small mammals like shrews, mice and bats. It involves a state of inactivity in which the rates of respiration and heartbeat are far below normal and regulation of body temperature ceases so that the animal's body functions like that of a poikilotherm. Some large mammals, including bears, sleep through most of the winter, but although they are inactive, their body temperatures and metabolic rates remain high.

PROGRESS TEST 5

1. Name the four main types of environmental factor. **(1)**

2. How has Liebig's law of the minimum been modified in the light of current knowledge? **(4)**

3. What are the main points of Shelford's laws of tolerance? **(5)**

4. Explain the combined concept of limiting factors. **(8)**

5. What is the importance of light quality as an environmental factor? **(12)**

6. Explain what is meant by the term compensation point. **(15)**

7. What are the main differences between heliophytes and sciophytes? **(16)**

8. Why is photoperiodism important in limiting the distribution of plants? **(19)**

9. Give two examples of photoperiodic reactions in animals. **(20–23)**

10. What are cardinal temperatures? **(26)**

11. How may plants be damaged by excess cold? **(28)**

12. What use can be made of vernalisation in horticulture? **(30)**

13. How may growing seasons be delimited? **(33–35)**

14. What is aestivation? **(39)**

CHAPTER VI

Environmental Factors 2: Water, Wind, Soil and Topography

WATER AND PLANTS

1. The importance of water.

(*a*) *Structure.* Water forms a large percentage of the living tissue of all organisms. Between 40 and 60 per cent of the fresh weight of trees is composed of water, and in the case of herbaceous plants the amount may be as much as 90 per cent. The cells contain an aqueous matrix which keeps substances in the correct state for the metabolism to function.

(*b*) *Support.* Plants require water for support in their non-woody tissues. If the cells of these tissues have sufficient water in them they are rigid. The pressure created by the presence of water in the cell is called *turgor pressure* and the cell is said to become turgid. If insufficient water is available, the turgor pressure is reduced, the cell contents shrink and the cell becomes *plasmolysed* (*see* Fig. 22).

(*c*) *Transport.* Plants use water to transport materials round their bodies. Nutrients entering through the roots are moved to other parts of the plant as substances dissolved in water. Similarly, some of the carbohydrates manufactured in the leaves are transported to the non-photosynthetic tissues in this way.

(*d*) *Cooling.* The loss of water from the plant by evaporation cools the plant body and prevents overheating.

2. How water enters the plant. Most land plants take in water through their root systems although some simple plants like mosses and lichens can absorb moisture directly from the air. Water enters roots through minute hairs which grow in a zone about 6 mm behind the root tip. These root hairs increase the surface area available for absorption and are constantly renewed as the root grows through the soil. Roots may obtain water by two processes.

(*a*) *Osmosis.* This is the movement of water from a weak

(a)   (b)

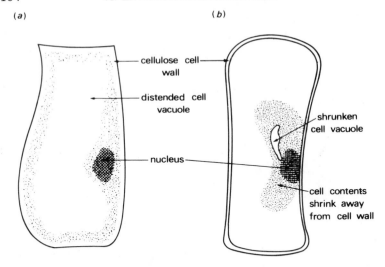

FIG. 22   *Turgid (a) and plasmolysed (b) plant cells.*

solution (water in the soil) to a strong solution (the cell contents) through a semipermeable membrane (the cell membranes of the root hairs). It is a passive process and accounts for most of the water entering plants.

(b) *Active uptake.* In this, water can be absorbed against an osmotic gradient. The process is not fully understood but it does require the expenditure of energy by the plant.

A plant which is growing needs to take in water in order to obtain nutrients from the soil. However, the rate at which water is taken into plants varies with several factors. The most important of these are the rate at which water is leaving the plant through evaporation, the acidity of the soil and temperature, because cell membranes are less permeable when they are cold.

3. The movement of water in plants. In the pteridophyta (the club-mosses, horsetails and ferns) and in the spermatophyta (the seed plants) water moves up the plant body in special conducting tissue called *xylem vessels.* The exact structure of those varies in the different groups of plants but generally they resemble very fine tubes. Water is drawn up through them partly by capillarity (the phenomenon that liquids rise in very narrow tubes against the force of gravity owing to surface tension) but mostly by the difference in vapour pressure between the foliage and the roots. This produces a constant flow of water through the plant.

In the simple plants which do not have xylem vessels water is transported round the plant body by osmosis.

4. How water leaves the plant. Most of the water which enters land plants is lost by evaporation. Only about 2 per cent of the water taken in by the roots is retained to build more plant materials. Water can leave plants by three methods.

(*a*) *Transpiration.* This is by far the most important mode of water loss. In the leaves water evaporates from the cell walls to intercellular spaces. From these it diffuses out to the atmosphere through minute pores in the leaf called *stomata* (singular = stoma) (*see* Fig. 23). The stomata are usually open in the daytime and closed at night. Their primary function is to allow the exchange of gases between the plant and the atmosphere.

(*b*) *Cuticular evaporation.* Some water may evaporate through the cuticle of the leaves and stems. Only a small amount of water is lost in this way (generally less than 10 per cent of total losses).

(*c*) *Guttation.* In very humid areas the losses of water by

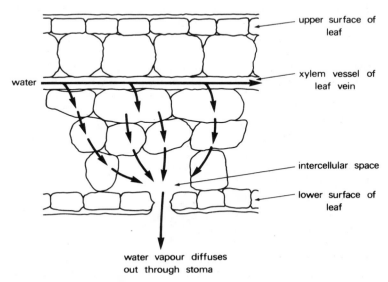

FIG. 23 *Diagrammatic section through a leaf to show transpiration.*

evaporation are too slight to allow more water to enter the roots. Some plants in these habitats have adapted by having pores at the end of the xylem vessels of the leaves. These pores, known as water stomata or *hydrathodes,* allow water to drip directly from the leaf.

5. Rates of water loss. The amount of water a plant needs, and consequently its environmental tolerance range, is determined largely by the rate at which it loses water. The rate is influenced not only by environmental conditions but also by the plant itself.

(*a*) *Environmental conditions.* Temperature, relative humidity and wind all effect the rate of evaporation and hence the amount of water lost from a plant.

(*b*) *Plant size and structure.*

(*i*) *Size of plant.* Obviously, a large plant usually needs more water than a small one. During a hot summer's day in Britain a mature oak tree would transpire about 675 litres of water whereas a maize plant would transpire only about $2\frac{1}{2}$ litres.

(*ii*) *Size of leaves.* Generally in humid areas, which have low rates of evaporation, leaves are large to facilitate transpiration. For example, tropical hardwood trees frequently have leaves 45 cm long. Conversely, the leaves of plants in arid areas typically have very small leaves to reduce water loss.

(*iii*) *Number and size of stomata.* The density and size of stomata vary considerably between species. Transpiration is more efficient through a lot of small stomata than it is through a few large ones. In *dicotyledons*, that is plants with two seed leaves (cotyledons) and broad leaves, stomata are mostly on the under surface of the leaves. For instance, privet has about 700 stomata per mm^2 on the undersides of its leaves and about 200 stomata per mm^2 on the upper surfaces. *Monocotyledons,* that is plants with one seed leaf and narrow leaves, have stomata on both leaf surfaces. Iris plants have about 200 stomata per mm^2 all over their leaves.

Plants which are adapted to live in dry habitats usually have few stomata. In some arid zone plants the stomata open at night and close in the day to reduce transpiration losses.

6. The effects of water deficit and excess. If a terrestrial environment is excessively wet the soil becomes saturated. The main problem resulting from this is lack of aeration in the soil. Roots are deprived of oxygen for respiration and the soil becomes acid.

If water supplies are insufficient for the plant's requirements the cells are flaccid. The stomata close to reduce further water loss. This state of water stress, known as *incipient plasmolysis*, may continue for a relatively long time. If the deficit continues the plant will die. Most plants suffer a partial dehydration during the day when transpiration is taking place, but restore their water balance at night.

7. Transpiration efficiency. Different plant species require different amounts of water to make growth. The ratio between net productivity and water transpired is the transpiration efficiency of the plant. It is usually expressed as grams of water transpired for every gram of dry weight organic matter produced. Plant species may transpire between 200 and 1000 g water to produce 1 g of dry weight matter. For example, the transpiration efficiency of wheat is 507, that of potatoes 408, and that of millet, which is an arid crop, 250.

A knowledge of the transpiration efficiency of a crop enables farmers to decide whether the plant will flourish in a particular climate and whether irrigation is necessary to produce maximum yields.

INTERCEPTION

We have noted that the functioning of a terrestrial ecosystem depends on its water supply. Water in the soil is replenished by rainfall and condensation. However, the amount of precipitation which is available for use by plants (that is the effective precipitation) is determined by other environmental factors especially temperature, wind and the biotic factor of the vegetation itself.

8. Basic idea of interception. Some rainfall will be intercepted by the vegetation canopy before it reaches the ground. This water will evaporate straight back to the atmosphere without becoming available for plant growth. The amount of precipitation intercepted depends both on the type of vegetation present and on the type and duration of rainfall.

9. Interception and vegetation type. The percentage of rainfall intercepted is related directly to the surface area of the vegetation. Communities with large surface areas, such as forests, experience the greatest interception losses. In some cases these are as high as 90 per cent of the rainfall. Much lower rates of interception occur in single layer vegetation and in ecosystems where the plants are widely spaced.

Interception is greater under coniferous trees than under broad leafed trees (*see* Fig. 24). This is possibly because the freer air movement through the conifers and the presence of smaller water droplets on the needle leaves encourage higher rates of evaporation.

Once the surface of the vegetation is wetted, water runs off to the ground by gravity. It occurs either by *direct throughfall* in the atmosphere or by *stem-flow* along the plant. The texture of the bark of trees influences the amount of water reaching the ground as stem-flow. It has been estimated that up to 15 per cent of precipitation may travel down the trunks of smooth-barked trees such as beech whereas only about 3 per cent travels down rough-barked species such as oak.

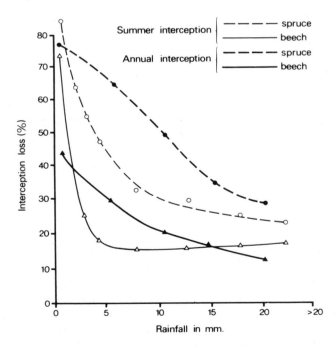

FIG. 24 *Interception losses from spruce and beech forests. (After Ward, 1967).*

10. Interception and precipitation type. Percentage interception losses are greater with light showers than with heavy downpours (*see* Fig. 24). Interception losses will gradually become less as rainfall continues because once the surface of the vegetation is saturated, direct throughfall and stem-flow occur.

If precipitation falls as snow, interception losses are generally increased. More evaporation can take place from the snow as it melts from the tree than from a similar amount of rainfall that would reach the soil faster.

11. The calculation of interception losses. The amount of precipitation intercepted can be calculated from the relationship:

$$I = R - R_g - S$$

where I is the amount of interception, R is the precipitation above the vegetation, R_g is the precipitation beneath the vegetation layer and S is stem-flow.

12. Fog-drip. Vegetation can increase the effectiveness of precipitation by acting as a condensation surface. Water droplets formed by condensation on the vegetation drip to the ground. This process of fog-drip or *horizontal interception* is especially important in areas with high atmospheric humidity, low cloud or persistent fog. In the coastal areas of northern California, fog is present for most of the year because of the cold Californian current offshore. In this location fog-drip contributes about 50 per cent of the total precipitation and allows the giant redwood (sequoia) forests to grow. The redwood trees are confined to the narrow coastal belt where fog-drip occurs.

THE PRECIPITATION BALANCE

13. The effectiveness of precipitation. This is usually expressed as the balance between evaporation and precipitation. Divisions between arid and humid areas are often made on the basis of where evaporation equals precipitation. However, it is difficult to determine the actual amount of evaporation taking place in an area as there are relatively few reliable records from which calculations can be made. Consequently, workers have favoured the use of the ratio between precipitation and temperature as an expression of the degree of aridity of a location. This approach is limited by the assumption that evaporation is a function of temperature alone.

14. Evapotranspiration. This is the water lost by evaporation from the surfaces of an ecosystem plus that transpired by the vegetation. Four variables control the rate of evapotranspiration.

(*a*) *Energy supply*. The energy required for evaporation is derived mainly from sunlight and is the ultimate factor determining water loss from the ecosystem. The reflectivity or *albedo* of the vegetation surface influences the amount of sunlight energy absorbed. Pine forests which have dull surfaces, may absorb up to 12 per cent more energy than grasslands which are more reflective.

(*b*) *Air movement*. Wind removes the water vapour and prevents the atmosphere becoming saturated. The removal of water vapour allows more evaporation to take place.

(*c*) *Vegetation type*. The aerial parts of the plants affect interception. The extent and type of root systems present influence the amount of water which can be taken into the plant bodies.

(*d*) *The amount of water in the root zone*. The rate of uptake of water by plants and hence its rate of loss by transpiration depends partly on the availability of water in the root zone.

15. The potential evapotranspiration index. In 1948 C.W. Thornthwaite, a climatologist, devised the concept of potential evapotranspiration (PE) in an attempt to overcome some of the problems involved with the calculation of precipitation balances. He defined PE as the total amount of moisture which would be evaporated from the soil and transpired from the vegetation of an area, if sufficient were available to meet all demands. Similarly Thornthwaite defined drought as a condition which arose when the amount of water needed for PE exceeds that available from the soil.

Thornthwaite constructed mathematical formulae to calculate values of PE from known temperature and light data. The relationship between precipitation and the calculated potential evapotranspiration index can then be expressed as a surplus or deficit of water in the soil. Figure 25 illustrates the relationship of mean monthly rainfall and potential evapotranspiration for Berkeley, California which has a mediterranean type of climate and for Seabrook, New Jersey which has a maritime temperate climate.

16. An assessment of the PE index. Thornthwaite's index has been criticised by some workers because it is complicated and gives a

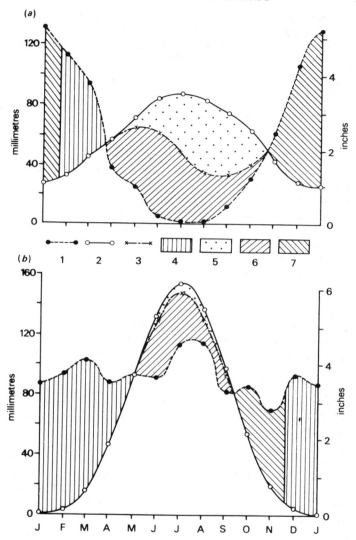

FIG. 25 *Examples of precipitation balances in (a) Berkeley, California and (b) Seabrook, New Jersey. Key: 1. Annual march of precipitation; 2. potential evapotranspiration; 3. actual evapotranspiration; 4. Water surplus; 5. water deficiency; 6. soil moisture use; 7. soil moisture recharge. (Adapted from Thornthwaite,* Geographical Review, *Vol. 38, 1948, with the permission of the American Geographical Society.)*

non-linear relationship between the variables used. Others have pointed out that temperatures tend to lag behind energy income and consequently there may be seasonal discrepancies if the PE index is worked out on a monthly basis. However, Thornthwaite's technique has been employed effectively by climatologists, soil scientists and hydrologists under a wide range of climatic conditions.

PLANTS' ADAPTATIONS TO EXTREMES

17. The adaptation of plants to a wide range of soil moisture conditions. This obvious specialisation led E. Warming, a botanist who worked at the beginning of this century, to distinguish broad groups of plants on the basis of their water tolerance. Although his groups are rather ill-defined the terminology is still used today. He recognised four main types of plant.

(*a*) *Hydrophytes*. These can grow in water and permanently waterlogged soils.

(*b*) *Halophytes*. These plants are specialised to cope with growth in saline environments such as salt marshes.

(*c*) *Xerophytes*. These are adapted for arid conditions.

(*d*) *Mesophytes*. This group contains unspecialised plants which tolerate moderate conditions between the environmental extremes.

18. Hydrophytes. The true hydrophytes are the angiosperms (= flowering plants) which have reverted to an aquatic environment, for example water lilies. However, the term is often applied to any plant of a wet habitat. The true hydrophytes may be floating or submerged. Their adaptations to an aquatic life include having no support tissue as it is unnecessary no cuticle, and having continuous air spaces in their bodies to give bouyancy. Gaseous exchange takes place in solution over the surface of the plant.

The flowers of hydrophytes are usually produced above the water to facilitate pollination. However, many species employ methods of vegetative reproduction instead, such as fragmentation. Lower plants like mosses, liverworts and algae are often restricted to damp habitats. Many of these, especially the mosses, have narrow tolerance ranges for water conditions. Consequently they can be used as environmental indicators.

19. Halophytes. Plants inhabiting saline environments must have mechanisms to deal with salt taken into their bodies. If salt were allowed to accumulate in the tissues the resultant high concentrations would impair the metabolism. Most halophytes secrete excess salt through special glands or hairs on their leaves. This process requires the expenditure of energy and reduces the net productivity of the plant.

Halophytes must also cope with the problems of physiological and actual drought. The presence of high concentrations of salt in the soil may prevent the plant taking in water by osmosis, thus inducing physiological drought. Halophytes in coastal salt marshes may suffer real drought during periods of low tide. The impact of these water stresses can be reduced by water storage in the plant. Many halophytes, such as *Salicornia* (the glasswort), have succulent tissue which acts as an environmental buffer.

20. Xerophytes. Plants with obvious adaptations for arid habitats are far fewer in number and more specialised than those of the other groups. The xerophytes can be divided into two broad groups; those which evade drought and those which endure drought.

(*a*) *Drought evaders*. These avoid dessication by adaptations in their lifecycle, morphology and physiology.

(*i*) *Ephemerals*. A large percentage of desert plants have very short lifecycles. They can grow from seed to the reproductive stage within a few weeks when water is available. Many of these ephemeral species synchronise their growth to rainfall availability by means of growth inhibitors in the seed. These have to be washed out by rain before germination starts.

(*ii*) *Succulents*. Perennial plants may evade drought by storing water in their tissues and by reducing water losses. Water may be stored in the leaves, as in agave, in the stems as in the Cactaceae and Euphorbiaceae or in the trunks of trees as in the boabab (bottle tree of East Africa).

In all of these cases the morphology (body-form) of the plant reduces transpiration losses. For example, stomata and intra-cellular spaces are few, leaves are reduced in size and the plants have very thick cuticles.

(*iii*) *Phraetophytes*. These are often known as the "water-spenders" because of their high transpiration rates and their ability to evade drought by finding and absorbing water. Their stategy is not to conserve water but to absorb as much as possible.

They have deep roots which reach *phraetic* or ground water. This is taken in efficiently by the high osmotic pressure of the roots. Phraetophytes such as the Palestine oak characteristically have small aerial growth compared with their extensive root systems.

(b) *Drought endurers.* Most plants cannot endure a marked reduction in their water content. For instance few higher plants can withstand more than 60 per cent desiccation. The capacity to withstand drought not only varies between species but also with the age of plant. Seeds and mature individuals can withstand greater water losses than seedlings can. Some desert angiosperms have developed a physiological endurance to dehydration. For instance, the cells of *Larrea* (the creosote bush of the Arizona desert) remain organised even when 70 per cent dehydrated. The leaves shrink as water is lost but regain their size and function when water becomes available.

Many of the lower plants, especially mosses and lichens, can withstand high percentage water losses.

21. The significance of xeromorphic features. There has been a tendency in the past to assume that the morphology of a plant must have some ecological significance. Workers have thought that direct causal relationships existed between form and environment. Ecologists are now aware that the morphology of an organ is not always a reliable guide to its function. Some species endure drought without possessing obvious morphological adaptations. Conversely, many species which do have xeromorphic features can transpire freely when water is available.

It has been noted that nearly all of the accepted xeromorphic features, such as reduced leaves and thick skins tend to develop in non-xeromorphic plants when they are subject to water stress or nutrient deficiency. Therefore, it is possible that these features are a result of water deficiency rather than an adaptation to it.

WATER AND ANIMALS

22. The importance of water. Animals require water for use in ways similar to those in plants. A large percentage of the bodies of all animals is composed of water. It is used for structure, for circulation and in the case of the homeotherms, for cooling. Animals exhibit a great range of tolerance to water conditions. In all cases the influence exerted by this factor is linked to the temperature of the environment. Water conditions become significant as a limit on distribution and function usually only in the case of deficit.

23. Damage caused by water deficit. This operates in conjunction with the effects of heat damage (*see* V, **39**). When an animal becomes dehydrated its body fluids thicken. Pumping denser blood causes the heart to be strained. Waste products are not diluted adequately and may become concentrated in the body. The animal's metabolism slows as substances become inactivated. In homeothermic animals lack of water impairs the body's temperature regulation mechanisms. If heat cannot be lost through evaporation the body temperature will rise and damage will be caused by excess heat.

The morphological, physiological and behavioural adaptations are related to those adopted to avoid damage to excess heat. These types of evolutionary developments are described in outline in V, **27** and are discussed in relation to habitats in XI.

WIND

Wind can exert a considerable influence as an environmental factor, both directly through abrasive action and indirectly through its control of other factors, especially temperature and water supply. It not only effects individual plants but also the species composition and function of entire communities.

24. Direct effects of wind action. Strong winds may restrict the growth of plants by inducing physical damage. Malformation of plant structure due to wind action is seen most often in exposed places like cliffs, mountain ridges and open plains. Persistent winds of high velocity may cause *windthrow* in trees which are rooted insecurely. The most dramatic damage occurs during tropical hurricanes in which practically all large plants are devastated.

In mid and high latitudes the combined action of wind and freezing rain causes the accumulation of ice on trees which may result in the collapse of the plant. In coastal areas the combination of wind and salt spray limits the growth of many plants as few can tolerate the salinity.

25. Indirect effects of wind action. Wind effects transpiration rates by removing water vapour from the vicinity of the plant thus allowing further evaporation to take place. This influence may be so strong that the plant's water balance is stressed even though soil moisture is plentiful. In these circumstances the vertical growth of the plant is limited by its ability to absorb and transport water upward to keep pace with the high transpiration rate. This results in the dwarf or elfin forms of trees characteristic of exposed habitats such as the Alaskan tundra, where mature birches and willows are only a few centimetres high.

The increased transpiration rates will be most pronounced on the windward side of the plant. This, plus the physical damage, retards the growth on this side producing *wind-shear*, which gives rise to an asymmetrical form as the leeward side of the plant has a faster growth rate.

High-velocity winds chill plants allowing cold to penetrate into the tissues. *Wind-chill* may distort growth or kill plants completely, restricting their distribution to sheltered habitats.

TOPOGRAPHY

It is well know that vegetation communities change with increasing height up mountains. Surface relief modifies all the climatic environmental factors. The fundamental effects of changing altitude are augmented by local variations in aspect and steepness of slope to produce a mosaic of ecosystems related closely to the landform.

26. Major effects of altitude. Topography may be sufficient to experience marked changes in temperature and humidity with height above sea level. Temperatures usually decrease with increasing altitude at a rate of about 0.65 °C per 100 m, which is slightly lower than the theoretical dry adiabatic lapse rate of 1 °C per 100 m.

Increasing altitude is frequently associated with increasing exposure and higher wind velocities. These act in conjunction with the decrease in temperature to influence humidity. The presence of high land induces orographic rainfall so ecosystems at high altitudes often receive more precipitation than low ones. These major modifications of the climatic factors produce zonations of ecosystems in relation to altitude. For example, on San Francisco Peak in Arizona (*see* Fig. 26) low altitude desert gives way to more humid forest formations, which in turn are replaced by alpine communities above the treeline.

In addition to the major alterations in climatic factors, altitude

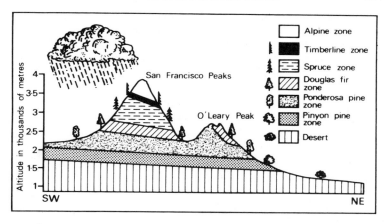

FIG. 26 *Vegetation zonation on the San Francisco Peaks, Arizona. (Modified from a diagram by CH. Merriam (1890), "Results from a biological survey of the San Francisco Mountain Region and the desert of the Little Colorado, Arizona,"* North American Fauna, No. 3.

also modifies light quality (*see* V, **11**). Increases in ultraviolet light received at high altitudes retard plant growth. These various climatic modifications operate as a major barrier to the migration and dispersal of plants and animals.

27. Aspect. Slope orientation is important in determining the amount of solar energy received at the surface. This has particular significance in mid and high latitudes where the sun is at a low angle especially during winter. Slopes facing the sun will be warmer and can support more biomass than those in perpetual shade. On a local scale, there are marked variations in species distributions between slopes of contrasting aspect. Generally, the warmer drier slopes support a more diverse flora. On a large scale, variations in slope aspect modify the altitudinal zonation of vegetation types. On the San Francisco Peaks, for example (*see* Fig. 26), the communities occur higher on the drier south-facing slopes than on the wetter north-facing slopes.

28. Angle of slope. The steepness of slope affects drainage and the stability of the surface. Generally, the steeper slopes support communities more tolerant of dry conditions than those of gentle gradients. In mountainous regions steep slopes are frequently associated with unstable deposits like screes. The frequent downslope movements of the surface materials prevent the development of closed communities thus keeping the ecosystem at an early seral stage.

EDAPHIC FACTORS

Environmental controls which are dependent on the soil are edaphic factors. Soil can be defined as the upper weathered layer of the earth's crust affected by plants and animals. This definition emphasises the close relationship between soil and organisms, both of which are affected by climate and topography. Indeed, it is extremely difficult to isolate the influence of edaphic factors because these are so interwoven with other aspects of the habitat. Many workers think that soil exerts its greatest affect where other environmental factors are extreme, for instance in very hot climates, very dry areas or on steep slopes.

The soil forms a complex part of the ecosystem and is inhabited by a wide range of organisms. The study of soils is known as *pedology*.

29. Soil as a medium for plant growth. Plants require four basic provisions from soil.

(*a*) *Anchorage for roots.* Plants must be anchored firmly in soil to withstand wind rock.

(*b*) *Water supply.* Most land plants absorb water through their roots. The soil must therefore provide adequate but not excessive amounts of water (*see* **2**).

(*c*) *Nutrient supply.* Organic and inorganic nutrients are present in soils due to the action of weathering and decomposition.

(*d*) *Air supply.* Soils must be aerated sufficiently to allow the respiration of roots and decomposer organisms.

Variations in supply of these requirements may lead to restrictions in the functioning and distribution of organisms and hence influence the structure of whole ecosystems. The edaphic essentials for plant growth are dependent on the physical and chemical characteristics of soils.

PHYSICAL PROPERTIES OF SOILS

Soils are composed of organic materials derived from the biotic part of the ecosystem, and inorganic materials derived from rocks by the process of weathering. The inorganic or mineral materials typically form about two-thirds of the volume of a soil and determine most of its physical characteristics.

30. Particle size. The amount and size of the mineral particles present depend on the type of rock which produced them and on the intensity of the weathering they have been subjected to. The particles can vary in size from large grains to minute specks which cannot be seen by the naked eye. They are usually grouped on the basis of size-classes or *fractions*. Various classification systems exist for particle sizes but two are used most frequently. These are the one devised by the International Society of Soil Science in 1926 and that devised by the US Department of Agriculture (*see* Table V).

TABLE V. CLASSIFICATION OF SOIL PARTICLES BY SIZE

Soil fraction	International system	US Department of Agriculture
Very coarse sand or fine gravel	– – –	2.00–1.00 mm
Coarse sand	2.0–0.2 mm	1.00–0.50 mm
Medium sand	– – –	0.50–0.25 mm
Fine sand	0.2–0.02 mm	0.25–0.10 mm
Coarse silt	– – –	0.10–0.05 mm
Silt	0.02–0.002 mm	0.05–0.002 mm
Clay	Less than 0.002 mm	Less than 0.002 mm

31. The clay fraction. Clay particles are important because they can hold water and nutrients in the soil. Some clay minerals can absorb three times their own volume of water, swelling when wet and shrinking as they dry. In addition, the clay particles adhere to each other making wet soil plastic and dry soil very hard. Humus combines with clay to form a clay-humus or *colloidal* complex which is relatively stable. This combination is not leached from the soil easily. These properties make clay the most important of the soil fractions and a major control on soil fertility.

32. Soil texture. The texture of a soil is determined by the proportions of the various soil fractions present. It not only affects ease of root penetration, aeration and drainage but also nutrient supply and soil temperature. Numerous textural types can be formed from the various proportions of soil fractions (*see* Fig. 27).

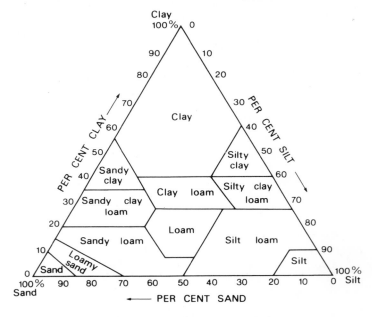

FIG. 27   *A soil texture triangle.*

33. Pore spaces. The number and size of spaces or pores in the soil will be a function of soil texture. Loose coarse sand will have large pores (macropores) although in total these may occupy only about 40 per cent of the soil volume. In contrast closely packed clays will have minute pores (micropores) but these may form 60 per cent of the volume of the soil.

34. The soil atmosphere. The air in the soil spaces contains a much higher proportion of carbon dioxide than is present in the free atmosphere because respiration by soil organisms is not compensated by photosynthesis. Gaseous exchange between the soil and the free atmosphere is largely dependent on pore spaces.

Soils with macropores will be better aerated and drained than those with micropores.

CHEMICAL PROPERTIES OF SOILS

35. Acidity and alkalinity. The acidity depends on the concentration of hydrogen *ions* (an ion is an electrically charged part of a molecule). The degree of acidity or pH is expressed as a negative logarithmic scale ranging from zero which is extremely acid, to 14 which is extremely alkaline (*see* Table VI).

TABLE VI. THE ACIDITY SCALE

Strongly acid	under 4.0
Moderately acid	5.0
Slightly acid	6.0
Neutral point	6.5
Slightly alkaline	7.0
Moderately alkaline	8.0
Strongly alkaline	over 9.0

Most soils in Britain have pH values of between 3.5 and 8.5. Those which approximate to the neutral condition are generally the most favourable for plant growth. However some plants are restricted either to acid or to alkaline soils. Species which grow only in alkaline soils, such as those developed on chalk downs, are called *calcicoles*, whereas species which grow only on acid soils, such as those of heathland (*see* XII), are called *calcifuges*.

Many interrelated factors influence the pH of a soil. The most important of these are the climate, through its effect on decomposition and leaching, the rock type from which the soil was derived and the vegetation, since this influences the rate of nutrient cycling and the chemical nature of the humus.

36. Humus. The type of humus is a product of climate and vegetation type. Plants which are nutrient-demanding produce organic debris rich in minerals, In well-aerated soils this encourages rapid decomposition resulting in a mild crumbly *mull* humus, which is conducive to a prolific and diverse soil fauna. Earthworms and other organisms transport the humus through the soil aiding aeration and promoting fertility.

Vegetation which absorbs few nutrients from the soil produces organic material deficient in minerals. It contains little calcium to balance the organic acids liberated during decomposition. In these conditions decomposition proceeds slowly, and acid *mor* humus is produced. If the soil is extremely acid organic matter may accumulate to form peat. Soils with mor humus have a sparse fauna and are infertile.

37. Inorganic mineral salts.

(*a*) *Anions and cations.* The soil solution contains mineral salts present as ions. These may be positively charged *cations* (known as bases) or negatively charged *anions*. Overall the clay–humus compounds are negatively charged so they attract and hold cations like those of calcium, sodium and magnesium. This association is loose as the cations can be released for plant growth and exchanged for others in a process of *base exchange.* Some bases are given up more easily than others. Generally the metallic ions, including potassium and sodium are released more readily than hydrogen. Consequently, soils become progressively more acid unless bases are replenished.

(*b*) *Acidity and nutrient avilability.* The acidity of soil affects the potential absorption of the mineral nutrients by plants in alkaline soils. Some minerals, such as copper, zinc and iron, may become insoluble so they are unobtainable to plants even though they are present in the soil. Conversely, iron and aluminium are highly soluble under acid conditions, and may attain toxic concentrations. Phosphorus combines with these two minerals forming insoluble compounds which cannot be used by plants. These reactions limit the distributions of certain plants.

38. Sodium salts.
High concentrations of salt (sodium chloride) accumulate in three main situations. Firstly, in coastal areas such as salt marshes and dune complexes; secondly, in areas of inland drainage; and thirdly, in arid areas where precipitation is insufficient to leach out the soluble sodium salts.

Soils with excess sodium salts are usually divided into saline and alkaline types. Saline soils have pH values below 8.6 and the amount of exchangeable sodium is less than 15 per cent. Alkaline soils have pH values between 8.6 and 10.0 and the amount of exchangeable sodium exceeds 15 per cent. Non-halophytes are precluded from these soils.

SOIL WATER

39. Water exists in the soil in three forms.

(a) *Hygroscopic.* This occurs as thin films held by surface tension round individual particles, this is sometimes known as *water of adhesion.* Hygroscopic water is generally unobtainable to plants.

(b) *Capillary.* The hygroscopic water attracts other molecules to form thicker layers of capillary water, which is capable of very slow movement from wetter to drier areas of the soil by the process of capillarity. The amount of hygroscopic and capillary water which can be held in the soil depends on its texture. Soils with fine fractions have a greater capacity for water retention than those with coarse fractions. Similarly, water can move further by capillarity in clays than it can in loams or sands. Capillary water is the main source of water for plants.

(c) *Gravitational.* When a soil contains the maximum amount of water which can be held by surface tension and cohesion on the particles, it is said to be at *field capacity.* Water in excess of this quantity will drain down through the soil by gravity (i.e. gravitational water). Its freedom of movement will vary with the size and number of the macropores in the soil. Obviously, sand will be better drained than loams or clays. If the gravitational water cannot drain away it will occupy the air spaces and the soil will be waterlogged. Gravitational water removes soil nutrients by leaching.

SOIL CLASSIFICATION

40. Soil profile. This is a vertical section through the soil. Each soil usually has stratified layers or *horizons* in its profile. In most cases at least three horizons are apparent. Conventionally these are assigned capital letters. The A horizon is subject to surface weathering and contains a relatively high percentage of organic matter. The C horizon is the weathered parent bedrock. The intermediate B horizon may display characteristics of the horizons above and below it. This threefold division has been refined by the recognition of an O (organic) horizon at the surface and an E (eluviated) horizon beneath the A layer.

Different soils possess contrasting development of their various horizons. This feature can be used to classify soils.

41. Classification systems. Many attempts have been made to produce a satisfactory classification of soils. However, it is extremely difficult to design a system which includes all of the soil variables. Two basic approaches are used most frequently.

(a) *The zonal system.* Soils are classified on the basis of their assumed origins in relation to broad climatic regions. *Zonal soils* have well developed profiles and reflect the influence of climate in their formation. Local factors may cause small-scale variations in the zonal type. These deviants, which are well developed are *intrazonal soils*. Soils which are developed poorly or are immature are called *azonal*.

Although the zonal system is simple and is useful for describing soils on a global basis, it has some disadvantages. One type of zonal soil may be found in more than one climate type. Also, soil profiles may not be a function of present climates. In some cases soils possess features inherited from previous climatic regimes. Lastly, the azonal soils may be the result of local inhibiting factors which should be acknowledged in the classification.

(b) *Systems based on soil properties.* Several systems have been devised to classify soils on the basis of their observable features rather than their assumed origins. For example, the US Department of Agriculture developed such a scheme, which uses new nomenclature throughout (*see* Table VII). This system suffers from the drawback that some of the diagnostic properties used for classification can be revealed only by detailed laboratory analysis. However, the scheme is becoming accepted widely because the primary soil orders are not associated with particular climatic or geographic environments.

Another example of the use of observable features was proposed by a British pedologist, B.W. Avery in 1956. This system emphasises the overall moisture status of the soil profile and considers the type of humus present. It is useful for agriculture in Britain but its application is limited.

Examples of soil types. The influence of climate on soil formation can be demonstrated by considering two examples of mature zonal soils.

(a) *Podsols.* These develop beneath many mid and high latitude coniferous forest and heathlands in regions where annual precipitation exceeds evapotranspiration. They are most frequent in the cool climates of the taiga areas (*see* XI). The predominant

Horizon colour:

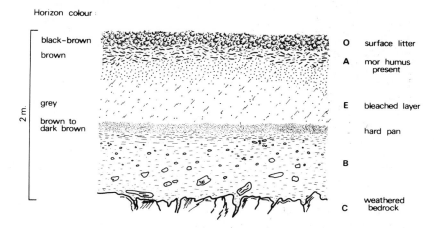

FIG. 28   *A generalised profile of a podsol.*

downward movement of water by gravity leaches soluble nutrients from the surface horizons. The infertile A horizon is acidic and may become bleached if leaching is severe (*see* Fig. 28). Nutrients may be redeposited in the B horizon sometimes forming an iron-rich hardpan which may impede drainage and root penetration. The climatic influences have little affect on the C horizon which returns its affinity to the bedrock.

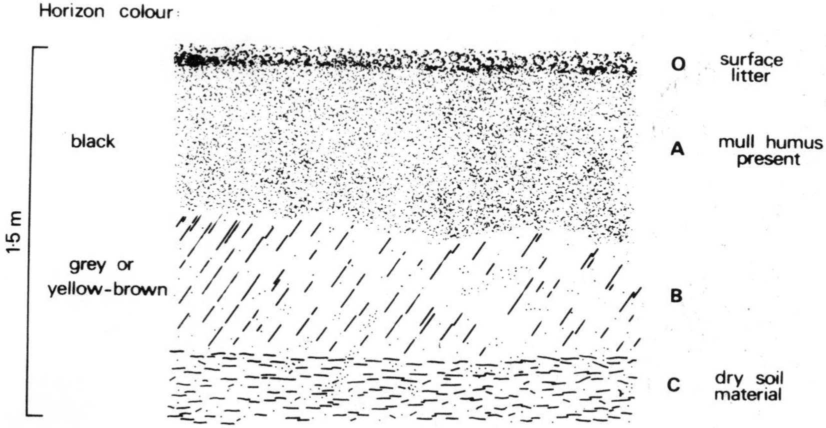

Horizon colour:

FIG. 29   *A generalised profile of a chernozem.*

(*b*) *Chernozems.* This soil type develops under steppe or prairie vegetation in areas where evapotranspiration exceeds precipitation. The annual net movement of nutrients in the soil solution is upward by capillary action. The deep A horizon (*see* Fig. 29) is black due to the presence of neutral humus and a base rich mineral matrix. The B horizon is often reduced in scale because of the great depth of the A horizon. This type of soil is often developed on loess (deposits of windblown silt) and is extremely fertile.

TABLE VII. MAJOR ORDERS OF THE USDA SOIL TAXONOMY

Order	Description
1. Entisols	Embryonic mineral soils
2. Vertisols	Disturbed and inverted clay soils
3. Inceptisols	Young soils with weakly developed horizons
4. Aridisols	Saline and alkaline soils of deserts
5. Mollisols	Soft soils with thick organic-rich surface layer
6. Spodosols	Leached acid soils with ashy B horizon
7. Alfisols	Leached basic or slightly acidic soils with clay-enriched B horizons
8. Ultisols	Deeply weathered, leached acid soils
9. Oxisols	Very deeply weathered, highly leached soils
10. Histosols	Organic soils

PROGRESS TEST 6

1. Why is water important for plants? (**1**)
2. What is capillarity? (**3**)
3. How may water leave plants? (**4**)
4. Which factors influence the rate of water loss from plants? (**5**)
5. What is the transpiration efficiency of a plant? (**7**)
6. How is the rate of interception influenced by precipitation type? (**10**)
7. Which variables control the rate of evapotranspiration? (**14**)
8. What are the main differences between hydrophytes and halophytes? (**17–19**)

9. Explain the phraetophyte strategy. **(20)**

10. What types of damage may be caused by water deficiency in the bodies of animals? **(23)**

11. What are the major effects of altitude on habitat conditions? **(26)**

12. Why is the clay fraction important in the soil? **(31)**

13. What are the main differences between calcicoles and calcifuges? **(35)**

14. What are the two types of humus? **(36)**

15. What is meant by the term base exchange? **(37)**

16. What are soil horizons? **(40)**

CHAPTER VII

The Ecology of Populations

POPULATION ATTRIBUTES

1. Definition. A population may be defined as a group of organisms of the same species occupying a particular place at a particular time. The population may be divided into *demes*, or local populations, which are groups of interbreeding organisms. The spatial and temporal boundaries of a population are vague unless the group is in a confined habitat such as an island. More usually the researcher delimits the physical boundaries of a population arbitrarily in order to study it.

2. Characteristics. A population of animals has various group attributes or characteristics which are not possessed by individuals. The basic feature of all populations is the size or *density*. This is extremely important because it influences ecosystem function. Density is affected by the parameters of *natality* (births), *mortality* (deaths) *immigration* and *emigration*. These parameters are in turn affected by aspects such as age structure, sex ratios and the pattern of distribution. The group attributes enable ecologists to recognise the organisation and structure of the population. The general genetic properties of a population form its *gene pool*. This is an important consideration in evolution and potential adaptation to changing environments (*see* VIII).

DENSITY

3. Measures of density. Density can be defined as the numbers of individuals per unit area or per unit volume. Occasionally measures of biomass are taken instead of counting individuals. Two types of population density are used.

(*a*) *Crude density.* This is a measure of the actual numbers or biomass of a population per unit space, for example 70 squirrels per hectare.

(b) *Ecological density.* This measure, sometimes known as economic density, considers numbers or biomass per habitat space. If, for example, the 70 squirrels in a hectare all lived in one small wood surrounded by fields, the ecological density would be much higher than the crude density. Ecological density avoids some of the inaccuracies that would derive from considering an unevenly dispersed population.

In many cases it is difficult to measure population densities accurately, that is to assess *absolute density.* Often measures of *relative density* are adequate whereby one area is known to have a denser population than another.

4. Density and trophic levels. The density of a population is related to the size of the organism and its position in the food chain. Since energy is dissipated via the food chains more energy will be available to support biomass at low trophic levels than at high trophic levels. Consequently, populations of herbivores will be denser than carnivores (*see* Table VIII). Related to this, the habitat range of organisms tends to increase towards the higher trophic levels of ecosystems. Predators need larger habitats in which to find adequate food than do herbivores.

Population density within a trophic level will be dependent partly on body size. Species with larger bodies will have lower densities than those with small bodies as the biomass resulting from the available energy flow is packaged into bigger units.

TABLE VIII. RANGE OF POPULATION DENSITY IN VARIOUS MAMMALS, STUDIED IN PREFERRED HABITATS, DATA FROM MOHR (1940)

	Crude density as biomass (kg/ha)
Deer (foliage-eater)	1.1–1.5
Squirrel (seed-fruit eater)	1.1–5.0
Fox (omnivore)	0.02–0.09
Weasel (carnivore)	0.001–0.008

5. Limits to density. The density of each species can vary within limits. The maximum density or *upper asymptote level* is the maximum number of animals which can be supported by the habitat. The *lower asymptote level* is the minimum number of animals required for breeding to compensate for deaths in the population. If density becomes very low reproduction rates decline. This is not only because finding a mate becomes harder but also because animals lose the urge to breed, particularly if the species relies on communal courtship displays as is the case with many birds.

CHANGES IN DENSITY

6. Increases in population size. Density increases by immigration and natality. The latter is the most important and is equivalent to birthrate except that it is a broader term covering all types of production excluding hatching and binary fissions.

(*a*) *Natality rate.* This may be expressed as a number of organisms born per female per unit time. The rates vary tremendously between animal groups. Some species breed once a year, some breed several times within a growing season and some breed almost continuously. For example, fish lay thousands of eggs, birds usually have clutch sizes of between one and 20, while mammals rarely have more than ten offspring at one time. Natality rates are related to position on the food chain, animals at low trophic levels (e.g. mice) having higher birthrates than those at high trophic levels (e.g. eagles).

(*b*) *Fertility and fecundity.* Fertility is the actual birthrate of a population whereas fecundity is the potential birthrate. For instance the fecundity rate for humans is one birth every 9–11 months per female of childbearing age, but the fertility rate may be only one birth every 8 years. Fecundity is inversely related to the amount of parental care a species devotes to its young. Fish, which give minimal care, have much higher fecundity rates than mammals, which look after their young for months or even years.

7. Decreases in population size. Density decreases by emigration and death. Mortality is the most significant cause. In the wild few animals die of old age. They are likely to be killed by predators, succumb to disease or starve. Therefore it is important to distinguish between *physiological longevity* which is the potential lifespan of an animal living in optimum conditions, and *ecological longevity* which is the average lifespan of an organism in its natural habitat.

Actual death rates can be correlated with age-groups in the

population to indicate life expectancy and probable causes of death. This can be done in two ways.

(a) *Life tables.* These were first used by Pearl and Parker in 1921 for Dall mountain sheep which graze on a reserve in Alaska where the main form of death is from predation by wolves. Details of numbers dying at each age, average mortality and survival rates are tabulated so that the average age at death and the chances of survival can be calculated.

(b) *Survival curves.* Similar information can be obtained from curves which plot the numbers of a population surviving through time (*see* Fig. 30). Numbers of survivors per thousand are plotted on a logarithmic scale against age of individuals. The graph indicates times of maximum death rate during a lifespan. A concave curve, as for example for oysters, shows that most of the population die young. A convex curve, as for example for man, shows that most of the population die old. A straight line, as in the case of hydra (a simple aquatic animal), shows a constant survival rate.

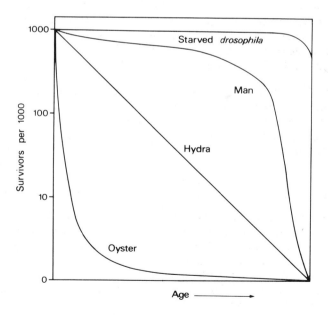

FIG. 30 *Survival curves for selected organisms.*

Some lifecycles have critical stages in them, for instance metamorphosis in insects and frogs. These stages involve high mortality rates and so produce a stepped line on a survival plot rather than a smooth curve.

AGE STRUCTURE

8. The percentage of organisms in a population at each age or stage in the lifecycle. These will influence natality and mortality; both of these parameters will affect population growth and stability.

(*a*) *Age pyramids.* The age structure of a population can be displayed as an age pyramid in which the proportion of the population in each age-group is shown (*see* Fig. 31). Age pyramids can be refined by dividing individuals into prereproductive, reproductive and postreproductive groups. The relative length of these stages varies greatly between species. Insects often have long prereproductive stages and no postreproductive stages; in contrast, man has a relatively long postreproductive stage.

(*b*) *Age–sex pyramids.* Each age-group in the pyramid can be divided into male and female organisms. This distinction is another useful indication of potential growth rates.

(*c*) *Age structure and stability.* If birth rates and deathrates are approximately equal the population size will be stable and will have an even age structure (*see* Fig. 31). If the birthrate exceeds the deathrate most individuals will be young and the population

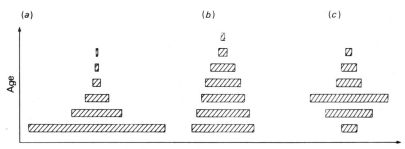

FIG. 31   *Basic types of age pyramids. (a) Expanding population; (b) stable population; (c) contracting population.*

will be expanding, so the pyramid will have a wide base. Conversely if the deathrate exceeds the birthrate the population will be contracting and the pyramid will have a narrow base.

Lotka (1925) has demonstrated that a population tends to develop a stable age distribution in a given environment. If habitat conditions are not changing the ratio of individuals in different age-groups remains relatively constant. Population age structures may be modified temporarily, by for example immigrants from an adjacent area, or permanently, by for example climatic changes.

GROWTH PATTERNS

9. Growth rates. The growth of a population depends on the relative rates of natality and mortality. High deathrates may be compensated by high birthrates. Conversely, low birthrates may not produce a declining population provided that deathrates are also low. The relationship between birthrates and death rates can be expressed in two ways.

(a) Crude growth rate $= \Delta N/\Delta t =$ change in the number of animals per change in unit time

(b) Specific growth rate $= \Delta N/N\Delta t =$ change in the number of organisms per change in unit time per organism

Growth rates are usually calculated for periods of at least a year because seasonal variations in population size can be misleading.

10. Patterns of increase. Studies of the way populations increase in size date from the work of Malthus in 1798. He put forward the idea that human populations would increase geometrically and would outstrip their food supplies. Notable advances in this field were due to the work of Verhulst in 1838 and to that of Pearl and Reed in 1920. These people recognised that populations tend to increase in characteristic patterns and they formulated equations to describe the various growth-forms.

(a) *The "J"-shaped curve.* When no environmental limits are operating the specific growth rate becomes constant and maximum for the prevailing conditions. In these circumstances the growth rate is a function of the inherent power of the population to increase in size, and is designated by the symbol r. This parameter is the *coefficient of population growth*. For short periods of time the population will increase exponentially by the relationship.

$$\frac{\Delta N}{\Delta t} = rN$$

where N is the number in the population and t is unit time. The rapid growth rate resulting from this relationship can be shown diagrammatically as a "J"-shaped curve (*see* Fig. 32). If the population size continued to increase indefinitely in this manner the carrying capacity of the habitat would be reached and the population would suffer an environmental disaster such as starvation. This would stop the population growth abruptly.

(*b*) *The "S"-shaped curve.* In natural ecosystems population growth rarely follows the "J"-shape pattern because environmental limits operate to slow down the rate of increase. Most populations increase their numbers by an "S"-shaped or sigmoid growth pattern (*see* Fig. 32). In this, growth takes place slowly at first, increasing to a maximum logarithmic stage, which is the steep section of the curve. As the carrying capacity of the habitat (upper asymptote level) is approached, environmental resistance operates to decrease the rate of growth. The growth curve flattens out before the maximum density is attained.

The sigmoid growth curve is equivalent to the relationship

$$\frac{\Delta N}{\Delta t} = rN \ \frac{(k-N)}{k}$$

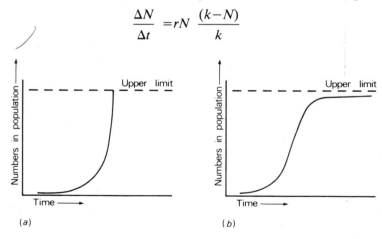

FIG. 32   *Population growth forms. (a) J-shaped (exponential); (b) S-shaped (sigmoid).*

in which k is a constant referring to the upper limit on population. As N becomes larger the growth rate decreases.

OSCILLATIONS

Once a population has grown to its upper asymptote level it rarely maintains equilibrium at this limit. Usually the population numbers oscillate in the range between minimum and maximum density. The amplitude of the oscillations varies between different organisms and different ecosystems. The most pronounced oscillations feature in simple ecosystems or where environmental conditions are changing. In contrast, more highly organised and complex ecosystems in stable environments exhibit smaller fluctations in their population numbers. The oscillations may occur on a seasonal or a periodic basis.

11. Seasonal oscillations. Variations in population density through the year are associated particularly with seasonal changes in weather and may be found in most parts of the world. Many animals in mid and high latitudes undergo great fluctuations in numbers between winter and summer. The greatest differences feature in species with short lifecycles, with limited breeding seasons or with pronounced seasonal dispersals.

Similar seasonal oscillations are found in tropical areas. In some cases these may be related to rainfall or to the periodic occurrence of food supplies such as fruit. Some complex seasonal oscillations occur for no apparent reason. For example, in aquatic ecosystems plankton numbers may suddenly multiply six-fold then decline again rapidly. No explanation has been determined for this.

12. Periodic oscillations. Fluctuations occurring over periods of several years are harder to explain. The oscillations may be minor, close to the asymptote density or they may be major variations.

(*a*) *The oscillations may be irregular and related to environmental conditions.* For example, the heron population of Britain declines markedly after each severe winter and recovers subsequently.

(*b*) *Some periodic oscillations appear to be regular and cyclic.* These may be due to intrinsic characteristics of the population. Controlled cyclic oscillations tend to fall within defined limits of density. For instance, the lemming populations of Scandinavia and Canada oscillate on a 4-year cycle. Peak numbers are followed by rapid declines in density due to migrations resulting in a high percentage mortality. Mice, voles and foxes are also subject to a 4-year population cycle. Other mammals, including the snowshoe hare and lynx of northern Canada experience peak densities every 9–10 years. Some insects, such as the desert migratory locust, oscillate in numbers over periods of about 40 years.

CAUSES OF OSCILLATIONS

An understanding of population fluctuations and their repercussions is necessary for a full interpretation of ecosystem dynamics. Basically, the oscillations may result either from environmental factors or from interactions within the ecosystem. There are four main theories on these aspects.

13. Meteorological causes. Attempts to relate periodic oscillations to climate have been unsuccessful despite some synchronisation of cycles in different ecosystems. It has been found that although peaks of abundance may occur simultaneously over wide areas, peaks in the same species in different regions do not always coincide. A climatic reason for these differences could not be found.

Previously workers thought that the 10-year cycle of many mammals could be explained by sunspot activity which induced major weather changes. However it is now thought that no correlation exists between these factors.

14. Random theory. In 1954, C. Cole suggested that what appear to be regular population oscillations are in fact random events resulting from the coincidence of random variations in complex environmental factors. However, L.B. Keith (1963) made a detailed statistical analysis of northern bird and mammal cycles. He concluded that the cycles were definitely non-random.

15. Population factors. Intrinsic population controls may act in conjunction with environmental factors. A population will increase until the environment becomes limiting. Thereafter it will decline until the effect of the limiting factors decreases. Thus oscillations would be generated by simple over population. The recovery of numbers back to peak density would be related to the intrinsic rate of natural increase. This will determine the length of the cycle.

These cycles may be augmented or generated by predator–prey interactions. For instance, the lemming forms the chief food for the snowy owl. When lemming densities decline most owls either starve or migrate, so causing a decline in their densities.

16. Nutrient recovery theory. Ecologists have observed that most population cycles are inherent characteristics of whole ecosystems rather than just single populations. Changes in density relate through food chains and involve changes in nutrient cycling. Schultz developed a "nutrient recovery hypothesis" in 1969 to explain population cycles in the tundra (*see* XI). He speculated that grazing by lemmings ties up and reduces the availability of mineral nutrients, especially phosphorus, the following year. The herbage available to the lemmings becomes low in nutrient content and the survival of the young is greatly reduced. By the fourth year the old lemmings die, liberating nutrients for plant growth. The subsequent recovery of vegetation enables the ecosystem to support a high density of lemmings again.

POPULATION REGULATION

17. Definition. Population regulation can be defined as the tendency for a population to return to its equilibrium size. Population stability is the tendency to remain at a constant size rather than to oscillate. It is a distinct advantage for a species to minimise its population oscillations, for if either asymptote level were reached the species would be in danger of extinction. In nature the extremes of population density are seldom attained. A great variety of mechanisms operate to restrict population growth.

18. Types of control. Charles Darwin realised in his work on the origin of species (1859) that although most organisms produce large numbers of offspring few survive. Overall, populations tend to remain within certain limits. Darwin recognised that this stability was a result of regulation by factors such as food supply, climate, predation and disease. It is current practice to divide these controls into two groups.

(*a*) *Density-independent.* These are all the aspects which operate irrespective of population density. For terrestrial habitats, they include temperature, solar radiation, wind, precipitation and, in the case of aquatic habitats, the physical and chemical properties of water. Density-independent controls tend to cause shifts in the upper asymptote levels of populations.

(*b*) *Density-dependent.* The operation of these controls is modified by population density. The group includes all the biotic interaction of organisms, such as predation, competition and social behaviour. Density-dependent controls tend towards equilibrium.

In practice it is often difficult to distinguish between these two groups as they are interrelated. For instance, climate influences food supply, which influences birthrate which affects density which may determine aspects of social behaviour that influence breeding.

19. The importance of different controls. The relative importance of the various controls on population growth has been debated at great length by ecologists. Three schools of thought have predominated in the past.

(*a*) From his work with birds in 1933, Nicholson deduced that most populations are balanced and stable. He assumed that fluctuations are relatively restricted and will be created by density-dependent controls. These would encourage growth after decline, and decline after growth.

(*b*) In 1954 Andrewatha and Birch suggested that since populations do become extinct, many must be potentially unstable. From this they deduced that fluctuations are irregular and are a function of climatic variations. Minor changes in climate would alter resource availability (e.g. food) and hence the carrying capacity of the habitat.

(*c*) Wynne-Edwards, working in 1962, recognised that density-dependent factors are important but he emphasised the role of food as the ultimate control. He thought that populations always maintained themselves close to the optimum level which food supply could support theoretically.

(*d*) Current opinion is that the relative importance of the factors varies with the type of habitat and organism involved: in low-diversity or physically stressed environments the populations tend to be regulated by physical factors such as temperature, in high-diversity ecosystems or those in benign environments the biotic factors tend to be limiting.

COMPETITION

20. Competition between organisms. This is found in every ecosystem and is an important factor in encouraging or inhibiting population growth. It can occur between individuals of the same species (*intraspecific* competition) or it can occur between individuals of different species (*interspecific* competition).

Many interactions between organisms are not detrimental and many even lead to mutual benefit, however in competition both species suffer from the association. For example, moose and snowshoe hares compete for the same limited food supply in winter.

Organisms compete for resources and space. Competition will be most likely in cases where organisms are seeking the same resources within the same environmental tolerance ranges.

21. The ecological niche. The niche of a species is its function and role in the ecosystem. The habitat can be thought of as where the animal lives and the niche as what it does there. The niche concept is important as it defines the precise conditions which a species needs.

(*a*) The fundamental niche is the maximum niche a species can have when not constrained by competition from other species. In practice the fundamental niche cannot be determined for all aspects as some environmental factors cannot be measured.

(*b*) The realised niche is the smaller niche occupied under the biological constraints of natural ecosystems.

(*c*) The vacant niche is a role in a ecosystem which is not filled. This void could be exploited by invading species.

(*d*) Breadth of niche refers to the range of tolerance of a species. Those with wide tolerances for food, resources and habitat have wide niches, whereas those with narrow tolerances have narrow niches.

22. Competition and niches.

(*a*) *Overlapping niches.* Competition is most intense between species occupying identical or similar niches. This usually happens between allied members of the same genus although unrelated species may occupy the same niche in different ecosystems.

(*b*) *Separation of niches.* In most ecosystems there are many species with diverse roles so that there are several niches at each trophic level. This functional specialisation of species avoids direct competition and aids stability in the ecosystem by providing alternative pathways for energy and nutrients. However, many amimals exist in partial competition with others because their niches overlap.

(c) *Specialisation and competition.* Field naturalists have observed many cases of closely related species living together in the same habitat. This coexistence is usually achieved through a high degree of specialisation. For instance birds may have very narrow niches for breeding and feeding, thus avoiding competition. However, if conditions are adverse mortality is high. In contrast, most freshwater fish are unspecialised; their wide tolerance ranges involve potential competition. Many species, however, have plastic growth rates; if conditions are adverse the fish stop growing until the environment improves. This flexibility counteracts some of the effects of competition.

23. Competitive exclusion principle. Two species cannot coexist in the same ecosystem indefinitely if they occupy the same niche. This idea is sometimes known as *Gause's hypothesis* after the Russian biologist who worked on populations in the 1930s. It implies that competition between two species occupying the same niche leads to the entire displacement of one by the other. This exclusion brings about an ecological separation of closely related or otherwise similar species.

24. Competition and succession. In the early seral stages species have wide tolerance ranges. Communities are open and there is little competition between species. In late seral stages, the complex structure and function of the ecosystem involves more competition between organisms. Niches are narrower and organisms are more specialised.

25. A mathematical model of competition. We have seen that most population growth approximates to a sigmoid growth curve which is equivalent to the relationship:

$$\frac{\Delta N}{\Delta t} = rN \frac{(k-N)}{K}$$

where N is the number in the population, t is time, r is the coefficient of population growth (*see* **10**(*a*) and k is the environmental resistance (see **10**(*b*)). If two species are competing they affect the population growth of each other. In this case another term must be introduced to the equation. The niche of a species

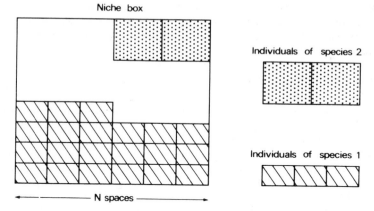

FIG. 33 *A model of interspecific competition.*

(species 1) can be thought of as a box (*see* Fig. 33) which can hold a certain number of individuals of the species, which can be represented by blocks. Another species (species 2) occupying the same niche may not need the same amount of "space" per individual in the box. For instance, it may be larger or require more food per organism. Therefore, a factor is needed to convert species 2 individuals into an equivalent number of species 1 individuals. This is done by the expression

$$N_1 = aN_2$$

(number of species 1 = conversion factor times number of species 2). From this it follows that the growth of species 1 in competition with species 2 will be:

$$\frac{\Delta N_1}{\Delta t} = r_1 N_1 \frac{(k_1 - N_1 - aN_2)}{k_1}$$

The same line of argument can be applied to species 2. A conversion factor is applied, so that:

$$N_2 = bN_1$$

(Number of species 2 = conversion factor times number of species 1). The equation for the population growth of species 2 in competition with species 1 will be:

$$\frac{\Delta N_2}{\Delta t} = r_2 N_2 \frac{(k_2 - N_2 - bN_1)}{k_2}$$

In nature it is extremely unlikely that the populations of the two species would be in balance. The species with the stronger competitive force would increase at the expense of the weaker one.

AGGREGATION

26. Effect on population densities. These can be affected by inherent tendencies for grouping or dispersal. Individuals in groups often experience a lower mortality rate during unfavourable periods or during attacks by predators than do isolated individuals. In addition, animals in groups may be able to modify microclimates, for example, by sheltering each other or by generating warmth.

Allee's principle states that undercrowding (lack of aggregation) can be as limiting on population growth as overcrowding. The advantages of aggregation have to be balanced against the increased competition for resources experienced by the animals in close proximity.

27. Types of grouping. Most populations exhibit varying degrees of clumping at some period. These aggregations may be due to the following:

    (*a*) responses to local habitat differences;
    (*b*) responses to daily and seasonal weather changes;
    (*c*) reproductive processes, forming colonies or family groups;
    (*d*) results of social attractions forming herds, flocks, etc.

Some sections of a population may live in groups while others are isolated. For example, although most lions live in family groups some are solitary.

DISPERSAL

28. Population dispersal. This is the random movement of individuals into or out of the population area. It supplements natality and mortality in shaping the population growth form. The gradual movements of individuals in search of food usually have little effect on densities as local emigrations are balanced by immigrations. Dispersal of young away from the parents operates to keep densities stable. Similarly, mass dispersals of individuals, such as lemmings, following a population peak prevents overcrowding. In all cases dispersal is influenced by barriers and the power of individuals to move (see X).

Migration is the movement of animals over long distances. These movements often involve high mortality thus reducing population numbers.

TERRITORIALITY

29. Definition. In many species of vertebrates and higher invertebrates, individuals, pairs or family groups confine their activities to a definite area or home range. If this area is defended it is called a territory.

The territorial spacing of the population can be the result of intraspecific competition for resources in short supply or of direct antagonism. The forces which bring about spacing of individuals are not as common as those favouring grouping. They occur most frequently in vertebrates and certain arthropods which have nest building and care of the young.

30. Types of territory.

(*a*) Some territories are temporary, such as those established and maintained by many birds for the duration of the breeding season. Other territories are permanent, such as those maintained by female voles.

(*b*) Different aspects of the territorial area may be defended. For example, in 1941 Nice classified bird territories in the following way:

(*i*) entire mating, feeding and breeding area defended;

(*ii*) mating and nesting area defended but not the feeding area;

(*iii*) mating area only defended;

(*iv*) nest area only defended;

(*v*) non-breeding areas defended.

This division is applicable to territories in general. Many territorial fish, amphibians and reptiles do not defend the feeding area at all.

(*c*) Within one species there may be various types of spacing behaviour. Some birds, such as robins, defend territories for breeding but aggregate in flocks during the winter. In other populations territorial behaviour is confined to a certain age-group or sex group. For instance, the young may aggregate but the adults isolate themselves.

31. Defence of the territory. In higher animals defence of the territory is through active behaviour although fighting over boundaries is kept to a minimum. The owners of territories usually announce their presence by song or display. Mammals often mark the borders of their territories with body scents (pheromones). Lower animals may also defend their territories actively but the use of chemical means is more frequent in these groups.

The size of the area defended can vary through the breeding season. The area defended by birds is often larger at the beginning of the nesting cycle, when the demand for food is great, than at the end.

32. The significance of territoriality.

(*a*) *The advantages of isolation.* The division of available habitat space into units encourages dispersal and so reduces intraspecific competition. Isolation of individuals can prevent overcrowding and the exhaustion of food supplies. These advantages of territoriality have to be weighed against those of cooperation and group action (*see* **26**). Which behaviour pattern prevails in a population depends on which gives the greatest long-term advantage.

(*b*) *Regulation of numbers.* It was thought originally that territoriality was a simple type of social behaviour which regulated numbers. Ecologists now realise that territoriality is a complex phenomenon. The significance of spacing varies with different groups of animals. In some it does seem to maintain the population at a level below the carrying capacity of the environment. Individuals which cannot establish territories do not breed.

The surplus organisms form a reservoir of individuals ready to take over territories if they become available through the death of their owners. This allows for a quick recovery if the breeding numbers of mortality occurs.

In some groups of animals territoriality does not seem to act as a population control. Many species of bird have "elastic" territories. That is, they vary in size according to the size or the population. If density becomes higher, available space is divided into smaller units.

(*c*) *Breeding success.* In birds territoriality may help to promote success in breeding rather than to regulate numbers. Evidence suggests that birds return to the same territories each year if breeding has been successful. If it has failed they establish a different territory. Defence of offspring and finding food are easier in a familiar area than in new areas which would be encountered if individuals wandered about.

(d) *Natural selection.* Territoriality operates as an evolutionary force by eliminating weak individuals. In territorial species, organisms which cannot establish and maintain territories do not breed so their characteristics are not passed on to the next generation.

PREDATION

33. Basic idea. Predation occurs when the members of one species eat those of another. It is a very important regulator of population sizes. If predators are removed from an ecosystem, the number of herbivores increases rapidly until they reach the upper asymptote level and the animals starve through overgrazing. In the long-term view the relationship between predators and prey leads to their mutual benefit. Predators gain food and the prey avoid overcrowding. In addition the predators are selective. They tend to take the diseased, old and weak individuals.

34. Predation and stability. If predators were to kill too many prey they would eliminate their food supply. Consequently, the relationship between the density of the two populations tends to become stabilised through time. This balance is achieved in two main ways.

(a) *Different birthrates.* Predators have lower birthrates than herbivores. This helps to keep their population numbers within the limits that can be supported by their prey. If for some reason the density of herbivores increases, some time will elapse before the carnivores can build their densities up to a comparable level. This sets off chain reactions. As the number of carnivores increases, so more predation takes place and the number of herbivores decreases so that fewer carnivores can be supported. In this way the system maintains equilibrium. The oscillations are usually small scale except where there have been sudden environmental changes.

(b) *Switching prey.* Food webs are usually complex. Predators typically eat several types of prey animal, which acts to stablise density fluctuations in the prey species. If one prey species increases in density the predator feeds on it more as it is more abundant. As its numbers decline, the predator switches to its other prey animals.

The Canadian goshawk, for example, eats grouse and hare. The numbers of each species taken vary according to their relative abundance. In this way population peaks in herbivores are dampened and the predators can maintain a stable population.

SOCIAL ORDERS

35. Dominance orders. In many animals which live in groups there is a clear order of dominance among individuals. This phenomenon is sometimes called the "pecking-order" because it was observed first in chickens. The dominance and subordinance relationships may be in a simple linear order or there may be complex rankings involving leaders and cooperative groups. These social organisations may operate to prevent overpopulation.

36. Significance for population regulation. If resources are in short supply the dominant members of a community will be the ones which survive. Subordinate members are the first to starve. In competition for territories or mates the high-ranking individuals succeed whereas the subordinates do not breed. For example, foxes live in family groups of one male with several females. Unless resources are abundant it is only the dominant female fox who breeds.

The presence of dominance in a group tends to eliminate some individuals from reproduction and therefore acts as a control on population growth.

37. Reactions to overcrowding in animals. This induces both psychological and physiological reactions which limit population growth. These changes can be very important population regulators. Most originate from disruptions in the pituitary–adrenal hormone systems. They can be of various types.

(a) *Urge to move.* Changes in the hormone concentrations of the bodies of overcrowded animals lead to restlessness and the urge to disperse. This occurs in the case of the lemming populations.

(b) *Decreased natality and maternal care.* In overcrowded mammal populations conception rate decreases. Spontaneous abortion and reabsorption of the embryo increases. Maternal care is lessened, for example the young may not be suckled.

(c) *Abnormal behaviour.* In severely overcrowded populations, dominance orders break down. Territories are not defended and the animals tend to become inactive. Feeding may cease and individuals may withdraw from social contact.

PROGRESS TEST 7

1. What are demes? **(1)**

2. What is the difference between crude density and ecological density? **(3)**

3. How does population density vary with trophic level? **(4)**

4. What is the difference between physiological longevity and ecological longevity? **(7)**

5. How is the specific growth rate of a population calculated? **(9)**

6. Describe the two main patterns of population increase. **(10)**

7. What are the possible causes of population oscillations? **(13–16)**

8. What is meant by the term population regulation? **(17)**

9. Give three examples of density dependent controls on population sizes. **(18)**

10. In which circumstances is interspecific competition likely to be most intense? **(22)**

11. What is the competitive exclusion principle? **(23)**

12. Why may animals aggregate? **(27)**

13. Define territoriality. **(29)**

14. What is the ecological significance of territoriality? **(32)**

15. How does the relationship between predators and prey lead to population stability? **(34)**

16. How may animals react to overcrowding? **(37)**

CHAPTER VIII

Evolution

INTRODUCTION

Over one and a half million distinct types of organism exist today. Any detailed study of these reveals an amazing degree of suitability of morphology, physiology and behaviour to the role of life and habitat of the species. Some plants and animals are extremely specialised and have obvious adaptations for particular niches. Others can exploit a wide range of environments.

Until the eighteenth century there was a virtual absence of theories about the origin and adaptation of life. This was mainly due to the limited experience of individuals with the natural world. However, for the past 100 years it has been accepted generally that the tremendous diversity of life-forms has originated from one or a few types of simple organism. All present and extinct types have been derived from these by the process of organic evolution.

EARLY THEORIES ON THE ORIGINS OF LIFE

1. Spontaneous generation. The earliest attempts to explain the origins of life are attributable to the ancient Greeks, particularly to Aristotle (384–322 BC). They believed that it was possible for life to originate spontaneously from inanimate material and that one type of organism could be transformed into another type abruptly.

2. Divine creation. Early christian teaching attributed all life forms to acts of creation by God. Types of plants and animals were thought to be immutable. The world was presumed to be perfect at the beginning. Any slight changes which may have taken place were viewed as deteriorations which must have occurred within the 6000 years since the supposed date of the creation. Similarly, the biblical legend of a universal flood was used to explain both the deposition and erosion of the earth's surface features.

THE DEVELOPMENT OF THEORIES OF EVOLUTION

The study of evolution encompasses geology and nearly all branches of biology. In addition, evolutionary theories have been influenced by theology. Modern ideas on evolution could develop only with parallel development in these other fields.

3. The contribution from geology.

(a) *William Smith (1769–1839).* It was customary in the eighteenth century to regard fossils as "wrecks of a former creation", that is of life prior to the flood of Noah. Smith, a practical geologist, recognised that the kinds of fossils changed as one went from lower to higher strata (layers) in the rock. He found that these changes were consistent across large areas so that stratified rocks could be identified and traced over long distances by means of the fossils they contained.

(b) *Charles Lyell (1797–1875).* His studies of fossils in stratified rocks revealed the gradual replacement of extinct species with modern species. Lyell's familiarity with aquatic animals enabled him to demonstrate that fossil shells in limestone beds several hundred metres thick on the coast of Sicily were of species now living in the mediterranean. This showed that the earth's surface features must have altered radically. Lyell realised that these changes caused extinctions and changes in distribution patterns.

4. The contribution from taxonomy.
Many attempts at taxonomy (classification) of plants and animals had been made prior to the eighteenth century. The majority of these were clumsy and failed to appreciate natural relationships between organisms.

Linnaeus (1701–78) proposed a hierarchical system which has been adapted for modern biology. In this, each type of organism is denoted by two names; one is a specific name unique to that type or species, the other is a genus name shared by others of the same closely related group or genus (plural = genera). For example, the American mink and the European mink are of the same genus, but are different species being *Mustela vison* and *Mustela lutreola,* respectively.

In modern taxonomy, which is based on the Linnaean binomial system, the plant and animal kingdoms are each divided up into major groups called *phyla* (singular = phylum). Each phylum is subdivided into classes, then orders and families. Each family is divided into genera, which are divided into the species that are the basic units of classification. This hierarchical grouping allows biologists to recognise the basic relationships between different types of organisms which are essential for a clear understanding of evolution.

5. Eighteenth-century theories. In the eighteenth century people began to accept that species could change through time. Several theories of evolution were proposed, of which two are most important.

(a) *Buffon (1707–88)*. This French naturalist made the first clear statement of a belief in organic evolution in that he realised that all life had evolved from pre-existing forms. Although he retained a belief in divine creation he proposed that species could change through time. His studies of living things indicated that variation between individuals of the same species was random. Consequently, Buffon deduced that gradual changes in species were away from original perfection towards slow decline and final extinction. Contrary to theological teaching, he thought that the earth had existed for at least 70,000 years.

(b) *Lamarck (1744–1829)*. This French zoologist was greatly influenced by Buffon and extended his ideas. Lamarck proposed a theory of true evolution, that is of gradual adaptation by organisms to suit their environments and modes of life. He maintained that species of animals could be related to each other by descent along branch lines of a "family tree". Thus, organisms in the same taxonomic group were related by evolution.

Lamarck held the view that changes in species were induced by their needs as reflected in their activities. Parts of the body used a great deal tended to improve while those parts not used gradually weakened and became reduced. He supposed that these characteristics acquired during the lifetime of an animal were inherited by its offspring. These ideas of "use and misuse" and "inheritance of acquired characteristics" were reasonable for the time as no-one knew of the mechanism of heredity. Lamarck's theory attracted many followers because it seemed to account for adaptations. For example, the long neck of the giraffe was seen as an adaptation acquired by use, which enabled the animal to reach high foliage.

6. The origins of modern theories. Our modern theory of evolution comes from the studies of Charles Darwin and Alfred Wallace.

(*a*) *Charles Darwin (1809–82).* Darwin worked as a naturalist on the British naval vessel *Beagle* during its voyage round the world. In this time he collected information about species distributions, diversity and the occurrence of fossils. He found that the fauna of the Galapagos islands contained different species of animals on each island. The animals were obviously closely related to each other and might have been derived from the same ancestral stock that had originally colonised the islands from the mainland. Individual species had become suited to their particular niches. This discovery led him to formulate his theory of organic evolution by natural selection. Darwin accumulated a great deal of evidence to support his theory but was reluctant to publish his ideas.

(*b*) *Alfred Wallace (1823–1913).* Wallace, also an English naturalist, thought of the idea of natural selection independently as a result of his work in the East Indies. He communicated his views to Darwin, who then agreed to publish a joint paper with him; subsequently in 1859, Darwin produced the famous book *On the Origin of Species by Natural Selection.* Although this was intended as an abstract of a fuller work it presents detailed evidence for organic evolution.

7. An outline of Darwin's theory.

(*a*) Darwin realised that the key to evolution was variation. Organisms within the same species are slightly different from each other.

(*b*) From the work of Malthus (1798) Darwin knew that all populations have a great potential for increase. All but the top predators produce large numbers of offspring but few survive. This leads to "competition for survival".

(*c*) The individuals which survive to reproduce will be those having favourable characteristics. These features are thus selected as suitable for the environment and are passed on to the next generation. Darwin did not know the mechanism of inheritance but realised that the selected features must be transmitted from one generation to the next.

(*d*) The natural selection of suitable characteristics brings about changes in species. This usually occurs over very long time-spans and results in adaptations to specific niches.

VARIATION

8. Types of variation. The study of variation in organisms and its measurement is *biometry*. Variations may be of two types.

(*a*) *Continuous variation* as in features which can be measured such as height or weight. Populations may exhibit a continuous range of variation for one of these aspects but most organisms will be of average measurements. If sufficient individuals of a species are measured a graph showing the numbers of individuals having particular dimensions will be a symmetrical curve known as a *normal distribution curve* (*see* Fig. 34(*a*)).

(*b*) *Discontinuous variation* as in features which can have alternative characteristics such as flower or coat colour.

9. Genetic variation. The *genotype* is the genetic blueprint of an organism and controls its development. The genotypes of a population are the ultimate source of the range of variation possible. Within the gene pool (total genetic material of the population) new genotypes may emerge by:

(*a*) *Mutations.* The genetic blueprint is held in the nucleus on chromosomes which are threads of protein. These contain the genes which are particles of deoxyribonucleic acid (DNA), responsible for controlling protein synthesis. Both the chromosomes and the genes can mutate either spontaeously in a random fashion, or by reacting to environmental influences like changes in temperature and radiation. Small mutations usually produce small-scale variations in the populations. These differences may be advantageous. Large mutations give rise to great differences in organisms. These are usually at a disadvantage and often die.

(*b*) *Recombination of genes in reproduction.* Reproduction in all but the most simple organisms involves a special type of cell division known as *meiosis*. One function of this is to mix genetic material so that new combinations of features emerge giving new variety in the population. During meiosis chromosomes cross over, break and exchange material with each other, resulting in new combinations of genes.

Sexual reproduction produces variation by combining genetic material from two individuals. Organisms which reproduce asexually have less potential for variation in their populations.

(*c*) *Polyploidy.* The number of chromosomes present in the nucleus is constant for each species. Occasionally in plants, cells

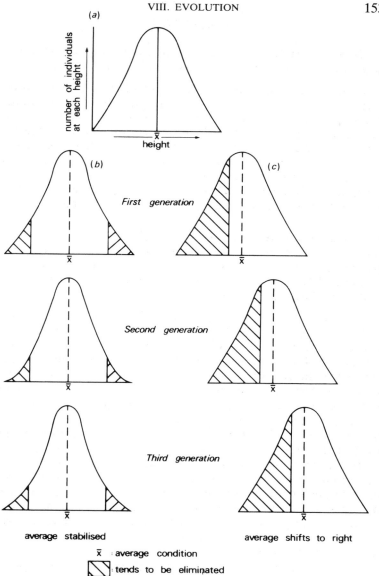

FIG. 34 *Natural selection acting on continuous variation (a) Normal distribution curve; (b) stable environment; (c) changing environment.*

fail to divide correctly with the result that a multiple of the usual number of chromosomes is present. This is known as polyploidy. Polyploid individuals are usually larger than their ordinary counterparts, which may give them an ecological advantage.

10. Environmental variation. The *phenotype* is the actual appearance of an organism. It is the result of control by the genotype modified by the environment. For example, if a plant has the blueprint for tallness but is grown in very adverse conditions it will be short.

Phenotypic plasticity is the range of expression possible for a genotype. Some features are more plastic (i.e. variable) than others. Sizes of flowers and leaves vary greatly according to environmental factors such as soil and light. In contrast, features like the arrangement of leaves on a stem and the number of petals in a flower are usually stable.

Sometimes it is difficult to determine which variation in a population is due to genotypes and which is due to phenotypes.

NATURAL SELECTION

11. On continuous variation.

(a) *Stable environment.* If habitat conditions are constant, the average condition for a feature such as height will be the most advantageous to an organism. Individuals which are exceptionally short or tall will tend to be at a disadvantage and will be eliminated from the population (*see* Fig. 34(*b*)). Their genotypes are removed from the gene pool so that the range of variation for the feature is less and a higher percentage of individuals have the average condition.

(b) *Changing environment.* If habitat conditions are altering, for example due to climatic change, the average condition for a feature may no longer be the best suited to the environment. One extreme of the normal distribution curve may be more favoured. For instance, exceptional tallness may become a great advantage. In this case more tall individuals are selected for survival. Short and medium-height organisms would be eliminated so that the average condition for the population shifts (*see* Fig. 34(*c*)). Thus the characteristics of the gene pool change. This is *genetic drift*.

12. On discontinuous variation. Natural selection in a changing environment acts to alter the proportion of alternative forms in a population. The percentage of organisms with the advantageous feature increases.

The British peppered moth, *Biston betularia*, exists in two colours, light and dark. The moth settles on the bark of trees during the daytime. If the bark is pale or covered in lichens the light form is camouflaged and is not eaten by birds. Conversely, if the trunks are dark, the dark form of the moth has the advantage. In 1848, populations of *Biston betularia* round Manchester were 99 per cent light-coloured. As air polution increased, lichens were killed and tree trunks became darker. By 1953 the colour balance had altered to 90 per cent dark-coloured as a result of selection pressure.

This shift in the proportion of alternative forms due to air pollution effects is known as *industrial melanism* and occurs in many species of moth in north-west Europe.

13. Adaptive radiation. As a species spreads out over its potential range by dispersal it may encounter new conditions. The environment factors vary over its distribution range. Natural selection operates to adapt the local populations to the prevailing conditions. In this way slight differences arise within the species. The local types, known as *ecotypes* or *demes*, exhibit morphological or physiological differences from each other but are still capable of interbreeding.

For instance, *Pinus sylvestris* (Scots pine) has 20 distinct ecotypes in its range from Scotland to Siberia yet all these are capable of producing fertile offspring if crossed.

SPECIATION

14. Speciation is the process of forming entirely new species. The modern definition of a species is a population capable of interbreeding and producing fertile offspring. Clearly, natural selection and adaptive radiation bring about changes within species but more is needed for speciation to occur. If the local races interbred the variation could be obliterated. The gene pools of the local populations need to be isolated so that their differences remain distinct until natural selection has altered them so much that they cannot interbreed with the rest of the original population. This can occur in three ways.

(*a*) *Allopatric isolation.* This is spatial separation, usually reinforced by barriers such as mountains, seas or adverse habitats.

Sections of populations could become isolated in this way during periods of climatic change, mountain building or island formation.

(*b*) *Sympatric isolation*. In this, the populations remain in the same area but some type of barrier operates to prevent interbreeding. For instance, the organisms may breed at different times of the year, occupy slightly different habitats or, in the case of plants, be self-fertilising.

If speciation results from sympatric isolation the new species have different niches to avoid direct competition. Sympatric isolation leads to specialisation within the ecosystem.

(*c*) *Temporal isolation*. One species may evolve into another through time. In this case of gradual, continual change it may not be possible to draw a clear division between two species although the ancestors and resultant forms are of distinct types.

15. Rates of speciation. The speed of speciation depends on the inherent variation of a population and environmental change.

(*a*) During periods when habitat conditions are stable little change will occur in organisms. Conversely, during periods of rapid change, in for example climate, speciation will also be rapid.

(*b*) Simple organisms, such as the blue-green algae, have simple genetic compositions. Often they have no method of gene exchange in the population as their reproduction may be asexual and may not involve meiosis. In this case variation in the population will be very limited and the potential for evolution will be small. Species like this will remain unchanged for long periods.

Complex organisms, such as the mammals, have complex genetic compositions and methods for gene exchange through the population. Variation between individuals can be greater and therefore evolution through natural selection can be rapid.

EXTINCTION

16. Extinction is an integral part of evolution. As new species evolve, others less suited to the environment become extinct. The process of extinction has several basic characteristics.

(*a*) Extinction is most likely to occur when environments are changing. Those species which cannot tolerate new conditions and which cannot adapt to them quickly enough will not survive.

(*b*) Extinction is most likely to occur in species with narrow tolerance limits. Organisms which become specialised for a particular mode of life lose the ability to adapt to changing conditions.

(c) Extinction is often preceded by the development of a relic community. The population of an endangered species becomes progressively smaller until only a few are left. These usually live in a habitat refuge, that is, in a place where conditions are still suitable. If environmental conditions improve generally for the species, the relic population may spread out again.

(d) Extinction of species and genera is common. Few orders and classes have become extinct; almost all the phyla have survived from their time of origin to the present.

(e) Types of organisms in which speciation is rapid, tend to have the greatest rates of species extinction. The turnover of birds and mammals is often much greater than in plants.

(f) The rate of extinction has been greatly accelerated by man's activities. Industrial techniques and agricultural practices have destroyed habitats and natural food supplies and caused pollution (see XIV). In the last 300 years man has completely destroyed over 200 species and has brought another 300 to the verge of extinction. This increased rate of extinction threatens ecosystem stability.

EVOLUTION ABOVE SPECIES LEVEL

17. Macroevolution. The evolution of the major groups of organisms is known as macroevolution, to distinguish it from the effects of variation, selection and speciation which collectively can be called *microevolution*, and the emergence of entirely new biological systems which is *megaevolution*. The essential features of microevolution are now well understood but the complex processes of macroevolution and megaevolution are still being researched. Many workers think that factors operating above the population level must be important. The history of the major changes in life are well documented from fossil evidence but the methods by which entirely new biological systems emerge are not understood.

(a) All macroevolution follows the acquisition of a new adaptation or the adoption of a new lifestyle. Within this new "life-zone" adaptive radiation takes place to produce many new species with diverging features.

(*b*) The new "life-zone" must be unoccupied by a strong competitor or the process of adaptive radiation cannot take place.

(*c*) Macroevolution involves sustained evolution trends. Certain characters tend to undergo progressive development so that organisms become more adapted to particular niches. Each adaptation limits the possibilities for future evolution.

(*d*) Adaptive radiation within the new "life-zone" tends to produce specialisations similar to those adopted by distantly related groups in similar habitats. This is *convergent evolution*. For example the cacti of the new world are very similar in morphology to the Euphorbiaceae of the old world.

18. An example of an evolution trend. A classic example of macroevolution is provided by the horse family, Equidae (*see* Fig. 35). The Eocene ancestor, *Hyracotherium* (*see* Table IX for

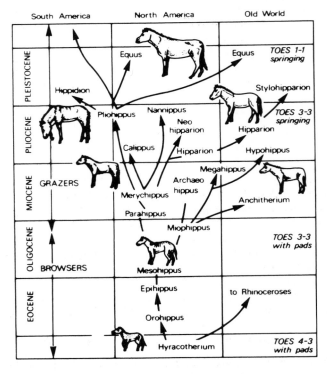

FIG. 35   *Evolution in horses from the Eocene to modern forms.*

TABLE IX.   THE GEOLOGICAL TIME-SCALE

Era	Period	Epoch	Date of commencement (millions of years ago)
	Quaternary	Holocene	10,000 years ago
		Pleistocene	2
Cainozoic	Tertiary	Pliocence	7
		Miocene	26
		Oligocene	38
		Eocene	55
		Palaeocene	65
Mesozoic	Cretaceous		135
	Jurassic		190
	Triassic		225
Palaeozoic	Permian		290
	Carboniferous		340
	Devonian		400
	Silurian		430
	Ordovician		500
	Cambrian		580
Precambrian		Oldest rock	3500
		Origin of earth	4500

geological time-scale) was a doglike browsing creature with padded feet. As habitat conditions changed from lush vegetation that could be browsed, to a drier grassland which necessitated grazing and speed of flight from predators, natural selection altered the population. Evolutionary changes involved the progressive elongation of the legs and head, and a reduction in the number of toes.

Many variations on this common theme resulted from adaptive radiation within the "life-zone". Although Fig. 35 is greatly simplified, it demonstrates the progressive, sustained development of the main features of the modern horse, *Equus*.

19. The origin of new biological systems. The emergence of completely new biological types, such as vertebrates from invertebrates, is rare. Only a few major types of biological system have developed in the history of life but almost all of them survive. The origin of these is the most important of evolutionary events but it is the least understood. The main features of megaevolution are:

(*a*) The breakthrough from one biological system to another, such as from aquatic to terrestrial life, always follows evolutionary experimentation through adaptive radiation. One trend in evolution manages to adapt to an entirely new mode of life.

(*b*) In order to survive, the breakthroughs must always be rapid and devoid of competition.

(*c*) Major breakthroughs are followed by periods of adaptive radiation.

(*d*) Each geological era has been characterised by one dominant life-form. Major environmental changes induce extinctions and the evolution of new systems.

THE COURSE OF EVOLUTION

20. The earliest forms of life. These can be recognised from the fossil record and are algae and very simple animals which existed in the Precambrian (*see* Fig. 36). Evolution from these primitive types was slow at first but accelerated under the impetus of environmental change.

(*a*) *The Cambrian.* The major groups of invertebrates (including sponges, corals, jellyfish, snails and trilobites which look like aquatic woodlice) had evolved by the end of the Cambrian. Similarly, the simple plant groups of thallophytes (algae and fungi), bryophytes (mosses and liverworts) and primitive types called psilophytes were present at this time.

(*b*) *The Silurian and Ordovician.* In these periods the first land plants emerged and the first vertebrates, which were jawless fish, developed. Fish became diverse and abundant in the *Devonian* as adaptive radiation took place. Scorpions, spiders and insects evolved soon after the appearance of vertebrates.

(*c*) *The Carboniferous.* By this period the amphibians and reptiles had evolved and plant life had developed to form the coal-producing forests. The reptiles were extremely successful and held ecological dominance for a long time, reaching their peak of diversity and superiority in the *Jurassic*. Primitive birds and mammals emerged at this time.

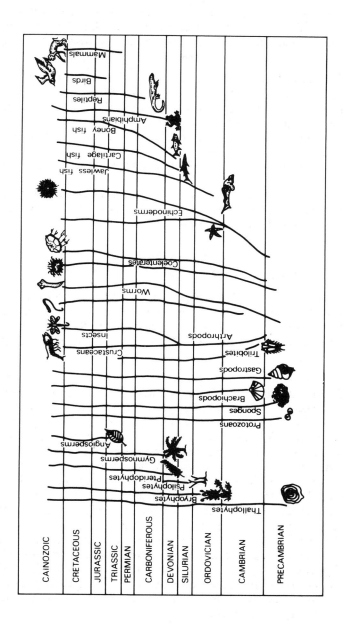

FIG. 36   The course of evolution.

(*d*) *The Cretaceous*. During this period the reptiles suddenly lost dominance, probably due to a climatic change. Many other groups declined with them. The early birds and mammals evolved rapidly to fill the ecological void, undergoing adaptive radiation without severe competition. The angiosperms (flowering plants) also became prolific by the middle Cretaceous.

PROGRESS TEST 8

1. Why was the development of modern taxonomy important to the study of evolution?   (**4**)
2. What was Lamarck's theory of evolution?   (**5**)
3. Give an outline of Darwin's theory of organic evolution. (**7**)
4. What is biometry?   (**8**)
5. How may new genotypes arise in a population?   (**9**)
6. What is meant by the term phenotypic plasticity?   (**10**)
7. What is adaptive radiation?   (**13**)
8. How can speciation occur?   (**14**)
9. What are the main factors influencing the rate of extinction?   (**16**)
10. What is macroevolution?   (**17**)
11. What are the main features of megaevolution?   (**19**)

CHAPTER IX

Strategies for Life

INTRODUCTION

Evolution of organisms has led to the development of various strategies for survival. These can be classified broadly into specialist and generalist modes of life.

1. Specialists. The evolutionary forces favouring specialisation are competition, sympatric speciation and extreme environments. It occurs most frequently in mature ecosystems of complex structure such as tropical rain-forests. In these cases, niches become extremely narrow. Life in extreme environments, like deserts, may not involve competition between organisms, but specialisations of morphology and physiology are required in order to survive.

Specialists have the advantage of being able to exploit particular resources more efficiently. The main disadvantage of the specialist strategy is that if conditions change the organisms cannot adapt to the new circumstances. Species evolved to exploit a narrow range of food and habitat types become extinct if these specialised requirements are not available.

2. An example of a specialist species. The hula bird (*Heteralocha acutirostris*) populations of New Zealand developed extreme specialisation of feeding technique. The beak of the male bird was short and tough whereas that of the female was long and slender. Each pair of hula birds cooperated in obtaining food. The male used his beak to break through the bark of certain trees after which the female penetrated the soft sub-bark layers to find insects. European settlers destroyed the tree–insect associations on which the hula birds relied. The beaks of this species proved to be so specialised that they could not adapt to an alternative food supply. The birds eventually became extinct in 1907.

3. Generalists. Species with wide ecological niches can exploit a variety of habitat types. Generalists tend to be most frequent in early seral stages in unstable ecosystems and in non-exacting habitats. The generalist strategy has the advantage that it retains the ability to adapt to changing environmental conditions and is therefore less prone to extinction. The main disadvantage is that it does not allow for extreme efficiency in the use of any one habitat or resource.

4. An example of a generalist species. The brown rat (*Rattus norvegicus*) is a good example of a generalist as it is present throughout Europe and has adapted to a wide range of food supplies. This species jumps, swims and digs very well. It has been introduced to many areas inadvertently by man. The rat's ability to exploit a wide variety of conditions has enabled it to destroy much of the native fauna of oceanic islands. For example, it was instrumental in the extinction of the dodo on Mauritius, and it helped to exterminate seven of the twelve species of bird on Lord Howe island (near eastern Australia).

SPECIES DIVERSITY

5. Within ecosystems. Most ecosystems contain a mixture of specialist and generalist species. In mature ecosystems, specialists tend to be more prolific than generalists because they have the advantage of being able to exploit a narrow niche efficiently. Characteristically, communities have a few species that are common, that is are represented by large numbers of individuals or biomass, and a large number of species that are rare at any given time and place. The common species are the *ecological dominants* and the rare species are the *incidentals.*

If a graph of the number of individuals of each species in an ecosystem is plotted the curve has a characteristic shape (*see* Fig. 37). The frequencies follow a *log-normal* distribution. Nearly all natural communities have this pattern.

The mixture of species strategies in an ecosystem is important. If environmental conditions change the ecological dominants may not survive. The many incidental species provide a reservoir of adaptability to evolve to suit new conditions.

6. Between ecosystems. Evolution has resulted in some eco-systems containing more species than others. The reason for this is not straightforward. Adverse environments like hot springs or early seral stages could not be expected to support many species because there has been insufficient time for evolution. The

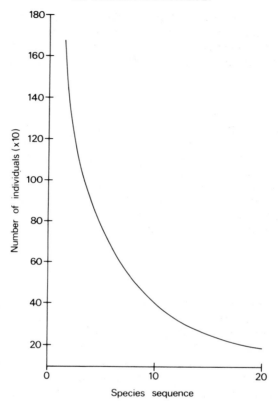

FIG. 37   *The log-normal distribution of species abundance.*

habitats are ephemeral and the chances of extinction are high. However, some places, like the arctic tundra (*see* XI) support few species even though the habitat is not ephemeral, and other apparently harsh places, like the Sonoran desert, support a rich diversity of plant life. Several theories have been proposed to explain these differences.

(*a*) *Global gradients.* There is an apparent gradient of species diversity from the tropics to the poles. Possibly the more complex structure of mature tropical ecosystems allows the development of more niches. Ecosystems in high latitudes are immature as they are still recovering from the last glaciation, so there has been insufficient time for evolution.

The disadvantage of this theory is that the gradient of diversity occurred also in remote geological periods without ice ages.

(b) Productivity. The number of species present is a function of energy flow through the system. The more productive ecosystems could, theoretically, have more species in them. However, some of the most productive habitats have very few species. The theoretical upper limit to the number of possible species is not likely to be reached.

(c) Habitat stress. The places with few species are the places with the highest probabilities for extinction. These could be ecosystems which are strongly seasonal, rare or ephemeral. Places where life is less stressful may not acquire new species by evolution any more quickly but the loss rate through extinction is less. Therefore, through time, species diversity accumulates.

ADAPTATIONS AGAINST PREDATION

Some of the most striking evolutionary adaptations involve protection from predation. Defence mechanisms can be behavioural, structural or physiological. Often a combination of these is employed.

7. Behavioural. This type of defence is used most often by social animals.

(a) Herding. A group of animals may scatter when attacked by a predator making it confused.

(b) Fighting. Many species will take active defence; for example, musk-oxen form a defensive ring to fight off attack by wolves.

(c) Evasive action. This may be simple speed of movement or may involve elaborate decoy movements; for instance, some species of moth make evasive leaps and dives when attacked by bats.

8. Structural. This type of defence can be used by solitrary or social animals.

(a) Increase in size. Predators usually kill animals smaller than themselves. Some species defend themselves by suddenly increasing their size. The puffer fish, for example swells up when attacked. This makes it difficult for other fish to swallow it.

(b) Camouflage. Many species rely on camouflage which can

be either colouration or shape. For instance the coat of the Arctic hare is brown in summer but white in winter to act as camouflage against the snow. Stick insects are not only green but also assume the shape of the vegetation they are on.

(c) *Colour warning.* Distinct colouration can warn predators that potential prey are toxic, taste bad or can sting. Examples of this defence include the distinct brown and blue colouration of the Panamian frog which warns of its poison mucus, the spots of the ladybird which warn predators of its unpalatability, and the stripes of wasps which give warning of their sting. Colouration warning assumes that predators learn from experience.

(d) *Mimicry.* Prey animals may mimic the features of species with warning colouration. This is a frequent method of reducing predation but it works only if the model is more frequent than the mimic. The mimicry can involve behaviour. For instance, the bordered bee hawk moth, which cannot sting, not only looks like a bumble bee but also mimics the flight and flower-visiting behaviour of this species.

(e) *Disruptive colouration.* The outline and orientation of an animal may be disguised by markings. The shape may be broken up by lines. Prominent markings, like false eyes towards the rear of the body, confuse the predator, especially if the real eyes and head are camouflaged.

(f) *Armour.* Defensive structures can be seen in many species such as the tortoise with the protective shell and the hedgehog with its spines.

9. Physiological. This type of defence can be used by solitary or social animals. It is frequently combined with colouration warning.

(a) *Stings.* Poison stings are characteristic of the insects. In some species the sting may be used many times but in others, such as the worker honeybee, the sting apparatus is left in the animal which has been stung. Often the act of stinging kills the honeybee itself.

(b) *Venom.* This features most frequently in reptiles, particularly snakes. Venom is injected through hollow fangs in the mouth.

(c) *Offensive discharges.* Some species rely on ejecting foul-smelling or toxic fluids. For instance, some termites exude a fluid from a gland in the head. The liquid evaporates to produce a repellent gas. Similarly, stink bugs defend themselves by giving off foul smells.

INTIMATE RELATIONSHIPS

10. Specialisation for life in intimate relationships. During evolution some species of plants and animals have become specialised for life in intimate associations with other species. These niches of close relationships are extremely important ecologically. The relationships can be of three types.

(*a*) *Parasitism*. An association of two organisms in which one, the parasite, exists at the expense of the other, the host.

(*b*) *Symbiosis*. An association in which two organisms exist in a close relationship for mutual benefit.

(*c*) *Commensalism*. An association of two organisms in which one derives benefit but the other is not affected. This type of relationship is relatively uncommon. Examples of it include the algae growing on a turtle's shell benefiting from the substrate provided, and the polychaete worm that lives in the shells of hermit crabs.

PARASITES

11. The niche. The term parasite comes from the Greek "*para*" meaning beside and "*sitos*" meaning food. The parasite lives off the tissue or food of its host rather than killing it outright as a predator would. It is in the interests of the parasite to keep the host alive so hyper-infestation, which would lead to death, is avoided. However, some parasites do cause diseases and decreased productivity. The parasite niche is a very specialised mode of life necessitating adaptations in morphology, physiology and lifecycles.

12. Types of parasite.

(*a*) Parasites can be grouped on the basis of where they inhabit their host.

(*i*) Ectoparasites live on the surface or in the superficial cavities of their host.

(*ii*) Endoparasites inhabit the internal tissues of the host. In the case of animals these include the alimentary canal and its associated organs.

(*b*) Parasites may be *facultative*, that is they can adopt another mode of life, or they may be *obligative,* that is they can exist only as parasites. Some species are parasitic only at certain stages of

their lifecycle. Generally the facultative parasites cause most
damage to their hosts as they have less incentive to keep the host
alive.

(c) Animal parasites are found in nearly all groups of the
animal kingdom. Some classes, such as the Hirudinea (leeches)
and the Cestoda (tapeworms) are entirely or largely parasitic.

Plant parasites are less diverse taxonomically than animal
parasites. Most are in the fungal groups. For this reason attention
is given to animal parasites in the rest of this section.

13. Characteristics of the lifestyle. The basic features of the niche
are similar for all parasites. This has resulted in *convergent
evolution*, that is similar environmental pressures have led to
similarities in morphology and lifecycle in different groups of
organisms.

(a) *Transfer between hosts.* In order to avoid hyper-infestation
and to continue life after the death of the host, the parasite must
move from one host to another. The period of transfer involves a
time of independent life.

(b) *Mortality.* Transfer between hosts involves high mortality
rates. These are offset by a great capacity for reproduction. Lifecycles are frequently elaborate and may involve *polyembryony,*
that is, they have many larval stages.

(c) *Alternating between hosts.* Many animal parasites have
lifecycles which involve alternate infestation of vertebrate and
invertebrate hosts. For example, the malaria parasite completes
part of its lifecycle in man and part in the mosquito.

14. Characteristics of morphology.

(a) *Degeneration.* The parasites, especially the endoparasites,
are adapted to a stable and sessile lifestyle. They do not require
sense organs so have reduced central nervous systems. Similarly,
muscles are reduced as movement is minimal. Because of these
trends the parasites are often called structurally degenerate.

(b) *Attachment.* Most parasites have methods of attaching
themselves to their hosts, for example, by suckers, and hooks.

(c) *Reproductive organs.* The necessity for a high reproductive
rate has led to the evolution of large and elaborate sex organs in
parasites. Many parasites are hermaphrodite in order to overcome the problems of finding a mate within or on a host.

(d) *Larvae.* The larval stages of animal parasites are very
different from the adult. Larvae are often adapted for movement

and are sensitive so that they can find a host. Adults are adapted to feeding and reproduction.

15. Advantages and disadvantages of parasitism. The main advantages of a parasitic lifestyle are that the adult inhabits a constant environment with an assured food supply. No energy is expended in procuring food or in sensing the environment so all productivity can be used for reproduction. The main disadvantages of the niche are that large numbers of offspring die on transfer between hosts and that the fate of the species is intimately linked to that of the host. In addition, the high incidence of self-fertilisation among parasites leads to decreased genetic variability and hence reduced chances of further adaptation.

16. The liver fluke. This is an example of an endoparasite.

(*a*) *Classification.* The liver fluke (*Fasciola hepatica*) is in the class Trematoda of the phylum Platyhelminthes, which are unsegmented flat worms.

(*b*) *Structure. Fasciola hepatica* has a leaf-shaped body approximately 2 cm long. Adults, which inhabit the bile ducts and livers of sheep and cattle, have two suckers for attachment (*see* Fig. 38). The oral sucker surrounds the mouth leading to a simple gut and digestive glands. Reproductive organs occupy most of the body. The species is hermaphrodite, the testes being near to the rear of the body and the ovary near to the middle. These organs connect with a common apperture (the gonopore). The rest of the anatomy is simple. There are no specialised sense organs and muscle tissue is minimal.

(*c*) *Lifecycle.* The liver fluke alternates between its vertebrate host (sheep or cattle) and an invertebrate host, a snail (*Limnaea truncatula*), which lives in damp places. *Fasciola hepatica* exhibits polyembryony (*see* Fig. 38). Increase in numbers occurs at each larval stage.

(*i*) Adult flukes produce vast quantities of fertilised eggs which pass into the gut of the body and out with the faeces. The eggs hatch to produce the first larval stage, the *miracidium*. This is mobile, sensitive and adapted to seek out and penetrate a second host. It is capable of following a mucilage stream left by a snail.

(*ii*) Once inside a snail's body the miracidium develops into a sporocyst which resembles a large sac. The next larval stage, *redia*, is formed by asexual reproduction within the sporocyst.

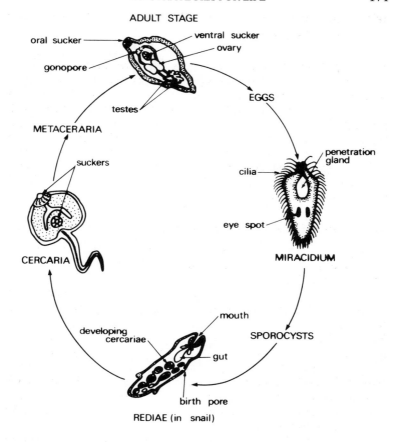

FIG. 38 *The lifecycle of the liver fluke (*Fasciola hepatica*).*

The rediae live inside the snail and give rise to the next stage, *cercaria.*

(*iii*) The cercariae, which are also mobile and sensitive, leave the snail. They form resistant cysts on vegetation. These cysts, or *metacerariae*, are quiescent and longlived. Eventually, they may be eaten by a sheep or cow. If so, they escape from their capsules and travel to the bile duct of the host to complete the lifecycle.

SYMBIOSIS

17. The niche. Many ecologists consider symbiosis to be a controlled or advanced form of parasitism. Examples of association for mutual benefit are prolific throughout the animal kingdom but symbiosis is probably more important ecologicaly in the plant kingdom. Many cases of plant symbiosis are critical to the functioning of ecosystems. During evolution many species have become so specialised for symbiotic life that they cannot exist alone.

18. Mycorrhiza. These are associations between fungi and roots in the soil. The fungi aid nutrient uptake in the roots and in return obtain carbohydrates from the root's food supply. The best-known examples are between tree roots and basidiomycetous fungi. Many trees will not grow without mycorrhiza. Mycorrhiza may be *ectotrophic*, in which case the fungal mass forms a sheath surrounding the root tips, or they may be *endotrophic*, in which case the fungal mass penetrates the root as well as surrounding it. In both types, the associations between species are varied. For instance. Scots pine (*Pinus sylvestris*) can have 119 different species of fungus in mycorrhizal associations. These fungi may associate with over 700 species of tree.

Roots which have mycorrhiza are shorter and more branched than other roots. This is mainly due to the presence of indole acetic acid (IAA) secreted by the fungi.

19. Root nodules. Associations between bacteria and roots form root nodules capable of nitrogen fixation. The bacteria derive food and shelter from the root whilst the plant gets nitrate from the association. Nodules are best known in legumes such as clover and peas. The bacterium in these is usually *Rhizobium radicola* or *R. leguminosarum*. Leguminous crops are important in agriculture because of their nitrate fixation. Many non-legume plants such as alder and sea buckthorn also bear nodules but less is known about these.

All nodules have limited lives. They are continually dying off and being renewed on the plant. Neither of the symbionts can reduce atmospheric nitrogen to nitrate on their own. In the nodule haemoglobin (the same red pigment as in human blood) aids the formation of ammonia (NH_3) from nitrogen, and then the production of nitrate (NO_4) by oxidation.

Root nodules are extremely important in the nitrogen cycle (*see*

II) because without this method of fixation the nitrate content of soil would be very low.

20. Lichens. These simple plants are associations between fungi and algae. Fungi obtain food from the algae while the algae gain shelter and nutrients from the fungi. Usually each lichen contains one species of each type of plant but multiple associations, containing several species of algae, do occur. Lichens are important colonisers of new areas and often form the initial seral stages of ecosystems. They are extremely tolerant of environmental extremes but they are very sensitive to air pollution (*see* XIV).

21. Animal–plant symbiosis. Many instances of this occur, three cases are given here as examples.

(*a*) *Cellulose digestion.* Bacteria in the guts of herbivores break down cellulose enabling the cell contents of vegetation to be digested. Bacteria gain warmth and food from the association.

(*b*) *Hydra.* The aquatic coelenterate, hydra, contains unicellular green algae in its body wall. These provide the hydra with oxygen and some food. In return the algae receive carbon dioxide and protection.

(*c*) *Ants and fungi.* Several types of tropical ant cultivate fungal gardens on decaying leaves in their nests. The fungi thrive on the decomposing vegetation and the ants gain access to the cellulose energy of the leaves.

22. Animal–animal symbiosis. Most examples of this type of symbiosis are loose associations rather than intimate relationships between two individuals. For instance oxpecker birds collect ectoparasites from rhinoceroses and antelopes. Cleaner fish perform the same duties for many species of large fish.

PROGRESS TEST 9

1. Which evolutionary forces favour specialisation?   (**1**)
2. What are the advantages of the generalist strategy?   (**3**)
3. Why are incidental species important in ecosystems?   (**5**)
4. What are the main forms of behavioural defence employed by animals as protection against predation?   (**7**)
5. What is meant by the term disruptive colouration?   (**8**)
6. What is the difference between endoparasites and ectoparasites?   (**12**)

7. Which morphological trends do most parasites have in common? **(14)**

8. What are the main advantages and disadvantages of the parasitic lifestyle? **(15)**

9. What is the function of mycorrhiza? **(18)**

10. Why are root nodules important? **(19)**

CHAPTER X

Migration and Distribution Patterns

INTRODUCTION

Most habitats can be occupied by many species. The presence and success of a species depends not only on its tolerance range and competitive ability but also on its capacity to spread from one area to another. Those species forming the biotic part of an ecosystem are the ones capable of living in it and which are available to do so. The spatial distribution of a species is a function of environmental controls acting in conjunction with the organism's inherent capacity to disperse. Examination of past and present migrations helps to explain the working of ecosystems and the anomalies in distribution patterns.

DEFINITIONS

Dispersal refers to short-distance movements as would take place between generations with the dispersal of offspring from the parents.

Migration refers to long-distance movements. Individuals may migrate during their lifespans. Alternatively, species can migrate by the accumulative dispersal movements of individuals over many generations. Migration may be temporary or permanent.

PLANT MIGRATION

1. Types of migration

(a) *Expansion of the distribution range.* The species may spread out from its centre of origin expanding into adjacent areas.

(b) *Change of distribution range.* The migration may involve a shift in the distribution, so that the range expands in one direction and contracts in the other. This occurs most frequently when climates change and species have to adjust their locations to the new conditions.

(*c*) *Contraction of the distribution range.* A species may reduce its distribution range in response to adverse conditions or competition. Although this is not technically migration it is often called *retreating migration.*

2. Speed of plant migration. In order to avoid deteriorating climates and to exploit new habitats before they become closed communities, a species must be able to migrate sufficiently quickly. The speed of migration depends on many factors.

(*a*) *Method of reproduction.* Plants may reproduce either by asexual or sexual means. Most asexual methods, such as by corms, tubers and bulbs, give very little movement between generations. Sexual methods, such as by seeds, have the potential for dispersal of offspring from the parents.

(*b*) *Method of seed dispersal.* Seed-producing plants contrast greatly in their potential migration rates. The method of seed dispersal determines the distance the species may move between each generation. Light windblown seeds may travel several kilometres from the parent plant. Similarly, seeds which are eaten by birds may be dispersed a long way. Heavy unpalatable seeds tend to remain close to the parent.

(*c*) *Germination rates.* The percentage of seeds that germinate varies between species and with environmental conditions. The stringency of the conditions needed for germination may limit migration as the seed is often the most vulnerable stage in the lifecycle.

(*d*) *Frequency of reproduction.* As plants can move only between generations, the rate of migration depends on how often the species produces offspring. Annual plants have the fastest potential rates of movement. In contrast, trees, which take many years to mature, are slow migrants.

(*e*) *Tolerance ranges.* The plant's ability to cope with changing environmental factors influences its capacity to migrate.

(*f*) *Competition.* During migration the species may have to compete with species already established in the new habitat. Rates of growth, shading and the efficiency of rooting systems are critical to the competitive ability of the species.

(*g*) *Human activity.* Man has tremendous influence on the distribution pattern of species. This is not only through disruption of ecosystems but also by intentional and accidental movement of species throughout the world (*see* XIV).

3. Barriers to plant migration. The migration of species may be halted by several types of barrier.

(a) *Environmental conditions.* The factors which limit species most frequently are temperature and water supply, although the other environmental factors are also extremely important. Ultra-violet light at high altitude may prevent the migration of a plant over a mountain range. Similarly, photoperiods may preclude a latitudinal shift in the distribution range.

(b) *Topography.* Changes in soil and climatic conditions encountered with changes in topography may act as barriers to migration.

(c) *Physical barriers.* Rivers, lakes and seas prevent the movement of species unless the dispersal mechanism of the offspring can cross the barrier. For example a species with windblown seeds could migrate across a lake, but a species which reproduces by bulbs could not.

(d) *Competition.* The existing flora of an area may exclude a new immigrant. The relative competitive abilities of the species determines which plants survive to inherit the ecosystem. Biotic factors usually present more barriers to migration than do environmental factors. Many species have extended their ranges once competition has been removed by human action.

(e) *Time.* If a species is slow to migrate it may be overtaken by environmental changes. Plants can be exterminated by climatic changes before they have time to colonise new areas.

4. Zones of sterility. Potential ranges of species are difficult to determine so it is not always clear whether a species is being limited in its distribution by a barrier. An indication of the operation of barriers can be seen in the condition of the adult plants. If environments are suboptimal each individual suffers reduced vigour.In adverse habitats the mature plants tend to exist in a vegetative state only, that is they fail to produce seed. Zones of sterile plants exist at the peripheries of many distribution ranges, indicating that the species is limited by environmental factors, particularly climate.

5. Good's theory of plant migration. Ronald Good's theory of plant migration states that if migration is induced by climatic change it can take place without evoking an evolutionary change in the species. A latitudinal shift in climate can be paralleled by a corresponding shift in the distribution range of a species. If a

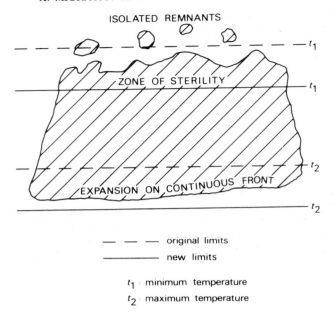

FIG. 39   *Diagram to illustrate Good's theory of plant migration.*

species can exist within a temperature regime between t_1 and t_2 (*see* Fig. 39) its distribution range will adjust as the temperature zone moves. This adjustment will occur by expansion along a continuous front in one direction and by irregular contraction in a zone of sterility in the opposite direction. Local pockets of the species may remain in favourable areas on the retreating edge.

6. The consequences of plant migration.

(*a*) *Change in distributions.* The most obvious consequence of migration is that the locations of species change. Since species migrate at different rates, communities will be disrupted. Some members will advance faster than others. New mixtures of species will come together to form ecosystems as the migrants mingle with the indigenous plants. The flora of the world has undergone many migrations in response to glaciations. These have resulted in the mixing of species from many areas.

(*b*) *Evolution.* Migrations often involve adaptive radiation. New habitat conditions will be encountered as the species alters its location. Natural selection operates on the gene pool to induce adaptive changes in the species.

(c) *Extinction.* The disruptions in ecosystems caused by large-scale migrations inevitably result in extinctions. Mass migrations induced by climatic change bring about the extinction of many species particularly those which are slow-moving, uncompetitive, or which have narrow ecological tolerance ranges.

ANIMAL MIGRATION

7. Introduction. Animal migration is a far more complex phenomenon than plant migration. The many types of movement found in the animal kingdom, such as seasonal changes in habitat and random wanderings, intergrade so that it is difficult to define animal dispersal and migration precisely. Since this gradation exists, many ecologists now consider all movements as types of migrations.

Migration occurs most frequently in the fishes, insects, birds and mammals. It has great importance for ecosystem function, evolution and the regulation of numbers.

8. Types of migration. A basic division can be made into temporary and permanent migrations.

(a) *Temporary migrations.* These may be seasonal or they may occur once in the lifetime of an individual. Temporary migration usually involves movement to a breeding or feeding area.

(b) *Permanent migrations.*

(i) *Random.* A species may extend its range by the accumulative dispersion of juveniles. Offspring tend to move away from parental areas in order to avoid overcrowding. The random wanderings of young alter the distribution pattern of the species. Extension of the species area can involve adaptive radiation.

(ii) *Induced.* A species may be forced to make a permanent migration by climatic change. Animals must have the ability to move to new areas to avoid climatic deterioration. Many permanent migrations have taken place in response to glaciations. These have caused disruption of communities, evolution and extinction.

Since animal movements intergrade some ecologists prefer to divide migrations into the following three types.

(c) *Exploratory migrations.* In these the animal moves beyond the limits of its familiar area but retains the ability to return to its home range.

(d) *Removal migrations.* Here the animal moves away from its known area usually to a comparable habitat without returning to its original location.

(e) *Return migrations.* These are movements to spatial units which have been visited previously. This type would include the three seasonal regular migrations between breeding grounds and overwintering grounds.

These types of migrations fulfil different functions for the species. They are not mutually exclusive; one species may exhibit all types of migration over a period of time.

9.   Seasonal temporary migrations. Seasonal migrations occur in many types of animal. In order to illustrate the principles involved examples are given here of both birds and mammals.

10.   European birds. Of the 450 species of birds in Europe, 175 are migratory. These vary in the extent of their movement, those nesting in northern and eastern Europe being more migratory than those in western Europe. This is mainly a reflection of climate. The birds overwinter in southern areas, including Africa and the far east, returning to Europe to breed in the summer.

(a) *Routes.* Birds follow definite flight paths when migrating. Approximately 150 species migrate on a general north-east to south-west route across Europe (*see* Fig. 40), while only about 25 species follow a north-west to south-east route.

(b) *Barriers.* There are two main barriers to migration, namely the Mediterranean sea and the Sahara desert. Large birds, such as the white stork (*Ciconia ciconia*), which could not sustain a long sea journey, cross the Mediterranean at its narrowest points, at Gibraltar and at the straights of Bosporus. Most birds traverse the Sahara by night, either crossing it in one flight or resting at oases. Some species avoid the desert by flying along the Nile valley.

(c) *Characteristics of flight.* Birds usually fly faster in migration than in normal flight. Starlings, for instance, which fly normally at about 48 km per hour attain speeds of over 72 km per hour during migration. Migrating birds fly in a determined way, on a straight course, often at high altitudes. For example, swifts migrate at about 2000 m, and geese at about 3000 m.

(d) *Impetus to migrate.* Most species migrate south before

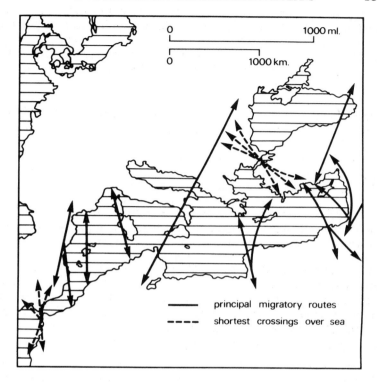

FIG. 40   *Generalised migration routes of European birds crossing the Mediterranean.*

conditions become adverse. Similarly return movements are unconnected with local food supplies in the overwintering grounds but seem to be geared to food supplies in the breeding areas which other birds are flying to.

This synchronisation is achieved by various stimuli. These include photoperiods, temperature and inherent hormonal rhythms. Some birds migrate at the same time each year whilst the activity of other species is modified by weather. The northward journey of the swallow shows a close correlation with the 9°C isotherm.

(*e*) *Navigation.* Many attempts have been made to discover how birds find their way during migration. Birds have a good visual memory of landmarks in their territories but successful migration depends on more than memory because young birds can travel to the migration area for the first time on their own. Suggestions for navigation methods include orientation by the position of stars or the sun, and magnetism.

11. Mammals. Many mammals are migratory but they have been studied less than the birds have, partly because a lot of the herding species, such as caribou and antelope, have been decimated by man. The migrations of two mammal types are described here.

(*a*) *Bison.* Vast herds of bison inhabited the plains of North America and Canada before the arrival of European settlers who slaughtered them. The bison made seasonal migrations of about 400 km on a north–south axis. Moving north in the spring for food and breeding, returning south in the winter to avoid the cold of high latitudes.

The bison herds used the same trails every year and commenced their journeys at regular times.

(*b*) *Baleen whales.* The migrations of baleen whales are between the feeding areas of polar seas and the breeding areas in the tropics. Baleen whales feed on plankton which is abundant in cold seas especially those round the south pole. During winter the whales migrate to avoid the adverse cold. They travel by regular routes, often close to shores. The young ones are born in the warm tropical seas but the adults do not feed there. Instead they use up the stores of blubber in their bodies.

12. Once-in-a-lifetime migrations. Temporary migration may take place only once in the lifetime of an individual for breeding. This is particularly prevalent in insects and fish.

(*a*) *Insects.* The activities of the migratory locust have been noted for centuries but it has been realised only fairly recently that other insects such as ladybirds and dragonflies also make migrations. Some of these movements are seasonal but the majority are once-in-a-lifetime migrations. Information collected over the last few decades has revealed that many species of moth and butterfly make such migrations over considerable distances.

For example, the European populations of the painted lady butterfly (*Vanessa cardui*) breed in the northern Sahara in the spring. The young butterflies fly north, some going as far as Scotland or north Germany. In the summer, when they are mature, the butterflies breed in these northern areas. When the young emerge, they fly back to the Sahara in the autumn. The insects travel along definite courses rather than just being blown by winds.

(b) *Fish.*

(i) *The north Atlantic eel (Anguila anguila).* This breeds only in the Sargasso sea, 800 km south-east of the Bermudas. The young migrate to north European coasts, taking nearly 3 years to swim across the Atlantic. They continue to move up rivers where they remain for between 7 and 19 years. After this time they retrace their journey to breed in the Sargasso sea. Once they have spawned, the adults die.

(ii) *The Atlantic salmon (Salmo salar).* This has freshwater breeding grounds. The young inhabit the swift-flowing rivers of western Eurasia and eastern North America. After growing for between 1 and 4 years, the young fish migrate downstream to spend the next three or four years as sea fish. When mature, they return to the rivers for breeding, then die. It has been proved that the adult salmon return to breed in the river in which they were spawned. This relocation is achieved by an extremely acute sense of smell.

SIGNIFICANCE OF TEMPORARY MIGRATION

13. Advantages of migration. It is very difficult to make generalisations or to develop a unifying theory about migration. Possibly, it has evolved in different types of animals for very different reasons.

(a) *Avoiding adverse conditions.* Most evidence concurs with the view that migration takes place in those species which survive in greater numbers if they leave breeding grounds in the non-breeding season than if they stay.

(b) *Avoiding competition.* Many of the wintering areas of species seem to be suitable for spending the summers in as well so it is difficult to see an advantage in moving away. Areas with adverse winters cannot support a large fauna all year. If there were no summer influx of migrants the food supply there would be wasted. It seems that many animals can raise more offspring in these ecological voids than they could in direct competition with other species in warmer climates. In this way, migration achieves the maximum use of resources.

(c) *Avoiding predators.* In several cases migrations are undertaken for safety. For instance, some species of duck move to islands where they can become flightless during moults without danger of attack by predators.

(d) *Regulation of numbers*. A high percentage mortality features in all migrations. It has been suggested that migratory movements help to limit populations. However, in most cases the same result could be achieved if the species remained in the breeding grounds since large numbers would die in the adverse season.

14. Origins of migration.

(a) *Climatic change*. The phenomenon of temporary migration was considered to have originated when animals were forced towards the equator by glaciations in high latitudes. Seasonal returns were thought to take place due to ancestral memory. The weakness of this theory was exposed by Mayr and Meise in 1930, who pointed out that glaciations were very complicated and that migrations were too diverse to be explained so simply.

(b) *Natural selection*. It is now accepted generally that migrations are a function of natural selection. Migratory movements confer some advantage to the species. Observations have shown that migratory habits may change gradually as would be expected if natural selection were operating.

(c) *Baker's theory*. In 1978, the animal ecologist R. Baker proposed the theory that all animals are migratory to a greater or lesser extent. With few exceptions, they move around to establish a home range. Wanderings may reveal seasonal differences in favourability of habitats which could lead to the development of movement on a seasonal basis. These movements may be accentuated by influences such as climatic change and continental drift. However, this theory does not explain the once-in-a-lifetime movements.

PERMANENT INTERCONTINENTAL MIGRATIONS

Permanent migrations of plants and animals have featured throughout the history of life on earth but those which took place in the Tertiary and Quaternary periods (*see* VIII, Table IX) are most relevant to modern ecosystems. An understanding of the basic features of these migrations helps to explain current distributions.

15. Fossil evidence.
The ancestry of living mammals can be traced by fossil evidence which proves that migrations occurred during

the evolution of particular groups. Mammals evolved quickly so that adaptive radiation took place in a relatively short time-span. This allows the history of individual types to be deduced with accuracy. The location of the oldest fossils of the type may be taken as the area of origin. Once this is established it is possible to determine the directions of dispersal and migration to other continents by examining successively more recent fossils. A fairly complete fossil record is required for this to be successful. For instance, as far as present evidence indicates the Proboscidea (elephant group) evolved first in Africa, but the early fossil record of this group is incomplete so it is possible that central Asia was the area of origin.

16. Evidence from present distributions. Modern distribution patterns show clearly that intercontinental migrations have taken place. In many cases related species are found in several continents. If the species of two continents are closely related, that is they are distinct but still similar, the migration has taken place relatively recently. If the two sets of species are related only distantly, evolution has had longer to operate, indicating that a relatively long time has elapsed since the migration.

Species of different continents show the greatest mutual affinities on migration routes. For example, the mammals of North America and South America are very similar either side of the isthmus of Panama.

17. Land bridges. During every glaciation vast amounts of water accumulated in ice-sheets, resulting in a lowering of sea level. At the peak of glaciation the sea level is estimated to have decreased by 100 m.

Areas of continental shelf were exposed to form links between continents. Most importantly, the area that is now the Bering Strait between Siberia and Alaska, formed a land bridge between Eurasia and North America. This connection facilitated the migration of species in both directions. The Bering bridge was formed and resubmerged many times during the Pleistocene glaciations.

18. Migration between North America and Eurasia. During the Tertiary these two continents possessed distinct faunas. The land bridges formed intermittently during the Pleistocene glaciations allowed an interchange of species to take place. The animals of

North America and Eurasia were even more similar in the Pleistocene than they are today because many extinctions have occurred since the faunas intermingled. The history of two families demonstrates the main pattern of migration.

19. Camels. Fossil evidence shows that camels evolved first in North America. The record is complete from their emergence in the Oligocene until their extinction 8000 years ago. The group evolved from small forms to ones of giant proportions (some over $3\frac{1}{2}$ m tall) in the early Pleistocene. Most of the American camels were adapted for life in warm climates. The migration of camels into Asia and their subsequent spread to their present distribution appears to have occurred suddenly. The modern bactrian camel is well adapted for life in cold deserts, having long hair resembling that of extinct mammals of glacial periods.

20. Elephants. The elephant group did not exist in North America before the Pleistocene. They originated and evolved in the old world. The first migrants to spread across the Bering bridge were cold-climate forms such as mammoths, indicating that the movement took place during the early glaciations. Once established within the new world the elephants evolved into new species suited to temperate climates. Northern American types included the world's largest elephants, for example the columbrian mammoth which existed between glaciations. All the American elephants became extinct in the middle and late Pleistocene.

21. Migration between North and South America. The land connection between the North and South American continents was formed in the late Pliocene and early Pleistocene. Until this time the fauna of the two areas had separate evolutionary histories. Evidence from fossils and current distributions shows that many species spread across the new link. For example, several species currently living in North America, such as the porcupine, armadillo and peccary spread northwards from South America. Other northward migrants, including the ground sloths and tapirs, suffered extinction in the late Pleistocene.

22. Pleistocene migrations in Europe. The periods between the Pleistocene glaciations (interglacials) have been of long duration and have been characterised by intervals of temperate climates. The alternation between cold and warm periods has caused successive northward and southward migrations of flora and fauna. During each migration the communities were disrupted and components of different ecosystems became mixed together.

23. Northward migrations. During the interglacials mixed broad-leafed deciduous forest spread northwards and eastwards over much of Europe as the climate became warmer. The diversity of the forest ecosystem decreased northward because of barriers to migration and time necessary for recolonisation to take place.

Many species of plants and animals achieved a much more extensive distribution than they have today. Evidence suggests that the interglacial periods were warmer than the climate has been since the last glaciation. For instance, species such as holly and beech, which appear to be limited by climate in Northern Europe at the present time had wider distributions during previous interglacials. In some cases species extended much further north than their present range. Examples include hazel and water chestnut, whose fossil remains have been found in Finland and Sweden up to 320 km north of the present limits.

24. Southward migrations. During the glaciations the fauna and flora of Europe migrated south so that species characteristic of high latitudes at the present time were prevalent in areas currently occupied by forest. Organisms now found in arctic and alpine areas inhabited much of Europe. Dwarf shrubs, such as least willow (*Salix herbacea*) and dwarf birch (*Betula nana*), which are typical components of the tundra (*see* XI), spread southwards as the ice-sheets advanced.

DISTRIBUTION PATTERNS

25. Basic patterns. Three main patterns in the distributions of oganisms can be identified. These may be applied to various groups on the taxonomic scale. For example, they can be used to describe the distribution of families, genera or species.

(*a*) *A continuous distribution.* This has no major gaps in its range.

(*b*) *A discontinuous or disjunct distribution.* This is made up of two or more separate areas with large distances between one location and another.

(*c*) *An endemic distribution.* This is confined to one area of limited extent, for example one country.

26. Continuous distributions. No species has a completely continuous cover. The difference between continuous and discontinuous distributions is somewhat arbitrary. Generally, if the space between the individuals within the habitable area are no greater than could be travelled by the dispersal offspring, the distribution is described as continuous.

Various types of continuous distribution can be identified.

(a) *Cosmopolitan organisms.* These have a world-wide distribution, and are types that must have very wide tolerance ranges. Only a few weeds and grasses are in this group, and have had their natural ranges extended by man.

(b) *Pantropic organisms.* These have continuous distributions throughout the tropics. For example, palms extend through the tropics and subtropics (*see* Fig. 41).

(c) *Circumpolar organisms.* These are found round the north or south pole, and the description should be confined to those inhabiting the polar latitudes and distributed in all longitudinal sectors. Many arctic species, such as the purple saxifrage (*Saxifraga oppositifolia*) have this distribution pattern.

(d) *Circumboreal (circumaustral) organisms.* These are distributed in the zones immediately adjacent to the polar regions (boreal in the north, austral in the south). The organism extends throughout all of the longitudinal sectors. The currant genus, *Ribes*, is an example of a plant with a circumboreal distribution.

27. Discontinuous distributions. There are several reasons why a distribution range can become discontinuous. The most important of these are migrations followed by extinction of part of the population, continental drift, and sudden long-distance dispersal as could occur if a seed was carried by a migrating bird or by a passenger on an aircraft.

There are four types of discontinuous distribution.

(a) *Diffuse.* Here the distribution is composed of numerous, small areas. For instance, mountain avens (*Dryas octopetala*) has its main location in high latitudes but is found also in many cool places such as coasts and mountains in western Europe.

(b) *Bipartite.* Here distribution has two separate locations within the same hemisphere. One area of location is usually much larger than the other. An example of this type is the plane genus (*Platanus spp.*) (*See* Fig. 41.)

(c) *Bipolar.* Here the distribution has two separate locations,

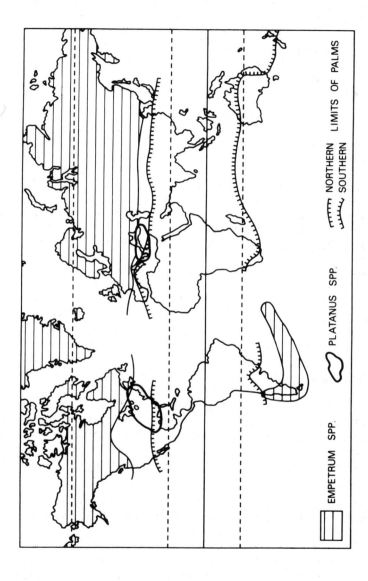

FIG. 41 *Examples of discontinuous and disjunct distributions (After Tivy (1971), using data compiled from R. Good (1953),* Geography of Flowering Plants, 2nd ed., *Longman Green, London).*

one in each hemisphere, as in the case of crowberry (*Empetrum*) (*see* Fig. 41).

(*d*) *Altitudinal.* Here the various areas of distribution are segregated into different altitudinal zones within the same region.

28. Endemic species. Endemism occurs when an organism is limited in its distribution by physical and ecological barriers. Endemic species greatly outnumber continuous ones.

Endemism can be of two types.

(*a*) *True endemics or youthful endemics.* These are of recent origin and have not dispersed far because of lack of opportunity or lack of time.

(*b*) *Epibiotics or old endemics.* These are older species that have had greater distribution ranges in the past, but have contracted into a restricted area. The small population of these usually lacks genetic variability rendering the species vulnerable to any changed circumstances.

29. Features of endemism.

(*a*) *Percentage and quality.* Endemism is examined in terms of its percentage, that is the percentage of the total flora which is endemic, and its quality, that is the number of major taxonomic groups that are endemic. A flora with a large number of endemic families would be of higher quality endemism than one with a very large number of endemic species all in the same family.

(*b*) *Broad and narrow.* Broad endemics occupy relatively large areas, such as parts of a continent, as in the case of cacao in the Amazon and the yellow pine in the western United States. Narrow endemics have very small distributions due to their limited ranges of tolerance.

30. Conditions favouring endemism. Endemism of high percentage and quality is associated with isolated places like oceanic islands and mountain tops. Habitats with a high degree of variety produce the most endemics. In these cases natural selection brings about separation but the new types cannot disperse to new locations. For example, in the Canary Islands (*see* Table X) the islands with the greatest altitude range, and hence the greatest range of habitat types, have the largest number of endemics. Variety is more important than size of island.

Old areas tend to have more endemics than newly formed places of comparable variety because evolution has had more time to progress in them.

TABLE X. ENDEMISM IN THE CANARY ISLANDS

Island	Area (hectares)	Maximum altitude (m)	Species endemic to one island	Species endemic to Canaries
Tenerife	1946	3715	81	233
Fuerteventura	1722	860	11	51
Gran Canaria	1376	1890	57	155
Lanzarote	741	684	8	434
Palma	706	2356	16	111
Gomera	378	1340	17	102
Hierro	278	1512	7	76

GEOGRAPHICAL ELEMENTS

31. Explanation. Within continents or within countries it is possible to recognise recurring patterns of distribution in flora and fauna. Groups of plants and animals have similar distribution ranges showing affinity to particular regions. These groups are known as geographical elements. The concept does not imply interrelationships at a functional level but does imply that the organisms have experienced a similar migratory history.

32. Early work on British elements.

(a) *Watson.* Geographical elements were recognised in the British flora and fauna by Watson in 1873. He divided the flora into six broad groups mainly on the basis of species distributions within the country. For example he described an Atlantic element which occurred chiefly in west and south-west Britain. This scheme had practical limitations as many species had no geographical bias within Britain.

(b) *Forbes.* In 1896, Forbes improved Watson's scheme by dividing the flora into five groups related to migration routes. These were the Iberian, the American, the Kentish, the Scandinavian or Boreal, and the Germanic. Forbes inferred that the groups of plants had migrated from these areas and were related to them ecologically, Forbes' scheme is now considered to be too simple.

33. Current work on British elements. The concept of geographical elements is still considered to be a useful aid to the understanding of past migrations and the origins of species although it is not always easy to identify migration routes or areas of affinity. This is particularly difficult in the case of ancient migrations. Many elaborate schemes for elements in the British flora have been devised. These include the ones suggested by Good (1953) and Matthews (1955). However, the simpler scheme proposed by Seddon in 1971 is widely accepted and has many advantages.

34. Seddon's scheme. This recognises ten geographical element species in Britain on the basis of distribution patterns within Europe.

(*a*) *Boreal.* These are distributed within the area of northern coniferous forests although they are not necessarily of forest habitat, for example marsh andromeda (*Andromeda prolifolia*).

(*b*) *Iberian.* These occur in south-west Europe and are widespread on the Iberian peninsular, for example wild madder (*Ribia peregrina*).

(*c*) *Mediterranean.* These are centred on the areas round the Mediterranean sea including North Africa, for example strawberry tree (*Arbutus unedo*).

(*d*) *South European.* These have a wide distribution throughout southern Europe, for example downy oak (*Quercus pubescens*).

(*e*) *Arctic.* These have their main location in the arctic or at high altitudes, for example mountain avens (*Dryas octopetala*).

(*f*) *Fennoscandian.* These occur mainly on the Scandinavian peninsula, for example bird's eye primrose (*Primula farinosa*).

(*g*) *Eu-Atlantic.* These feature close to Atlantic coasts, for example cross-leafed heather (*Erica tetralix*).

(*h*) *Sub-Atlantic.* These are distributed in western Europe occurring further inland than those of Eu-Atlantic group, for example foxglove (*Digitalis purpurea*).

(*i*) *Nemoral.* These occur to the south of the boreal zone in the area usually occupied by deciduous forests, for example mistletoe (*Viscum album*).

(*j*) *European-wide.* These have more extensive ranges throughout Europe than those of the nemoral group do. They extend north to Scandinavia and east to the Baltic, for example Scots pine (*Pinus sylvestris*).

RELICTS AND REFUGIA

Areas of discontinuous and old endemic distributions are often relicts of distributions that have been more extensive in the past. Study of these indicates the previous history of the organism.

35. Characteristics of relict species. Typically the relict species does not occupy all suitable habitats. It may have suffered local extinctions so that its distribution becomes fragmented and scattered. A contracting or relict species can be distinguished from a newly formed expanding one as the latter tends to spread out on a continuous front occupying all suitable habitats. A relict species does not necessarily become extinct. It may disperse out again if conditions become favourable.

36. Types of relicts. Relict areas can be classified according to their type and origin.

(*a*) *Migratory relicts.* Most migratory relicts are labelled by the time of the migration. For instance they may be the result of migrations during the Pleistocene glaciation in which case they could be labelled by the appropriate glaciation or interglacial period. Present areas of mountain avens in Britain are relicts of the last glaciation.

(*b*) *Evolutionary relicts.* A small proportion of a type may continue to exist after the extinction of the rest of the species. Evolutionary relicts include the maidenhair tree (*Gingko spp.*) which is different from all other conifers. Its relatives are known as fossils from the Permian. The coelacanth fish populations are relicts from the Cretaceous. This species was thought to be extinct but some living specimens have been found off the Indian coast since 1936 onwards.

37. Refuges. During periods of adverse conditions a species may retreat to small habitats where it can survive. It may remain in the refuge area for a long time possibly expanding out again when conditions become favourable.

Britain contains many examples of refuge sites. When the climate warmed after the last glaciation arctic-alpine species retreated northwards as more temperate species invaded. Some arctic-alpines have survived to the present day in open places, such as cliffs and gorges, which do not experience hot summers. Well-known refugia include Teesdale and Brean Down, Somerset.

PROGRESS TEST 10

1. What are the main types of plant migration? **(1)**
2. Which factors influence the speed of plant migration? **(2)**
3. What are the potential consequences of plant migration? **(6)**
4. What are the causes of permanent migrations of animal populations? **(8)**
5. What are the advantages of temporary migration for animal species? **(13)**
6. Explain how the habit of temporary migration may have originated. **(14)**
7. How do modern distribution patterns provide evidence of permanent intercontinental migrations? **(16)**
8. Why did the diversity of interglacial ecosystems in Europe decrease northwards? **(23)**
9. What are the three main types of distribution pattern? **(25–28)**
10. What is the difference between circumpolar and circumboreal distribution patterns? **(26)**
11. Explain the term "good-quality endemism". **(29)**
12. Which factors favour endemism? **(30)**
13. What are geographical elements? **(31)**
14. What is the nemoral element in the British flora? **(34)**
15. What is the difference between migratory and evolutionary relicts? **(36)**

CHAPTER XI

Major Natural Ecosystems

BIOLOGICAL DESERTS

1. Desert conditions. These are the result of deficiency in one or more of the essential requirements for life. Limits may be imposed by aridity, extremely hot or cold temperatures, toxic substances or high wind velocities. The most extensive deserts are the hot and cool deserts of the arid zones and the cold deserts or *tundra* of high latitudes in the northern hemisphere.

2. Basic features of deserts. Although the climates of hot and cold deserts contrast markedly there are basic similarities in these ecosystems.

(*a*) The harsh climates have limited and variable growing seasons. Organisms must be adapted to the adverse conditions. Specialisation in morphology and physiology is more abundant than in less vigorous habitats.

(*b*) Vegetation communities have a simple structure. Lack of tall growth results in a lack of complex layering. Communities are frequently open, that is, they have a discontinuous cover. Community composition varies in relation to local habitat factors, producing a mosaic of diverse vegetation units.

(*c*) Primary productivity is low so food chains are short and the total amount of biomass in the system is small.

(*d*) Soils are immature, lacking organic matter and the development of horizons.

(*e*) Desert ecosystems are instable. Productivity rates vary greatly in response to changes in environmental factors. Animal populations are subject to violent oscillations in numbers.

3. Arid zones.

(*a*) *Location.* Areas of extreme aridity occur either in the centres of continents or near western coasts within 30° of latitude from the equator (*see* Fig. 42). These areas cover approximately one-third of the world's land surface and include not only the hot

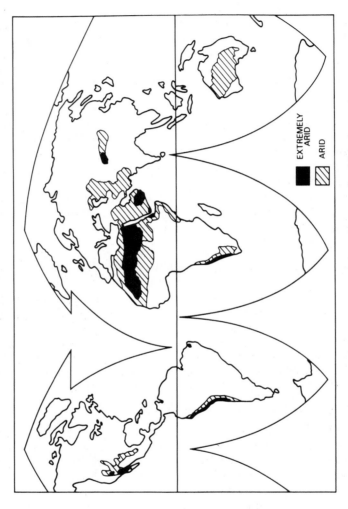

FIG. 42   *The world distribution of arid zones.*

EXTREMELY
ARID

ARID

deserts but also the cool and cold deserts of North America and Eurasia.

(b) *Environmental conditions.*

(i) *Precipitation.* Arid zones receive less than 100 mm of rainfall per annum and have annual potential evapotranspiration rates of about 1140 mm. Precipitation is irregular. Many months or even years may pass without any rain, followed by a heavy downpour; much of this storm water is lost by runoff and evaporation. Dew formation augments the water supply in many deserts.

(ii) *Temperature.* Arid areas experience great diurnal ranges of temperature because of the dry air and absence of cloud. For example, the difference between night and day temperatures in Arizona is typically about 56°C. In hot deserts maximum daytime temperatures may exceed 60°C. These extremes are accentuated by seasonal differences.

(iii) *Wind velocity.* The open habitat presents no barriers to decrease wind velocities. The force of the wind is often strong, causing damage to plants by mechanical injury and abrasion.

(iv) *Salinity.* Inland drainage features in many deserts. Evaporation from the surface leaves residues of salt which accumulate. In this way large areas of desert become extremely saline.

(v) *Microclimates.* The climatic constraints give very little potential for environmental modification by organisms, as for example by the sheltering effect of tall plants. In these circumstances very small-scale variations in temperature and humidity are important factors in determining the distribution of oganisms. Many species live beneath stones and boulders or in crevices in the soil where the microclimates are favourable.

4. Ecosystem function in arid zones.

(a) *Productivity.* Most work on desert ecosystems has been descriptive. The quantitative research which has been done indicates that primary productivity is very low, being less than 0.5 g/m^2/year. Actual amounts of productivity are a linear function of rainfall. Growth is limited to wet periods so the ecosystems undergo short, irregular bursts of productivity. The amount of biomass of the permanent standing crop is small, both at the autotroph and heterotroph levels.

(b) *Food chains.* Since the primary productivity is small the energy flow through the ecosystem will be limited. Trophic levels and system components are few. However, there are many interrelationships so food webs may be complex. Usually desert animals are not specialised feeders because they cannot afford to be dependent on one food type. The euryphagic habit acts as an environmental buffer; the animals can tolerate a wide range of food types and so exploit all energy sources available.

Variations in primary productivity are passed quickly through the food chains causing heterotroph populations to oscillate violently. This reflects the inherent instability of the system.

(*c*) *Nutrient cycling*. Desert ecosystems are deficient in nutrients so the amounts cycled are low. Even in the most fertile places usable nutrients are confined to the top 10 cm of the soil. The rate of cycling is slow because many of the plants are perennial and many of the animals are long-lived because it is necessary to complete lifecycles over several seasons. This longevity locks up nutrients for long periods.

A large percentage of the flora is adapted to cope with nutrient deficiency. Many are legumes capable of nitrate fixation in root nodules. Others, such as the podocarps, can fix nitrate in non-symbiotic associations with bacteria. Species which shed their leaves draw nutrients back into the stems before the leaves are lost.

5. Evolution in arid zones. Desert ecosystems are of great antiquity. Current arid zones are thought to have originated in the late Tertiary following climatic change induced by orogenesis (mountain building). Evolution over long periods in desert conditions has selected similar adaptations in morphology and physiology in unrelated organisms in different places.

Lack of competition and freedom from predators in deserts has allowed many ancient groups of organisms to survive. Few families are endemic to deserts. Most arid zone plants and animals are more specialised members of groups found also in less severe habitats.

6. Desert autotrophs. The main xeromorphic features are described in relation to the environmental factors of water and temperature in VI. In addition, the arid zone plants have two main metabolic adaptations. These occur in a high percentage of the vegetation in areas receiving less than 100 mm of rainfall per annum.

(*a*) *Crassulacean acid metabolism (CAM)*. This was discovered first in the Crassulaceae (stonecrop family). It enables carbon dioxide to be stored in malic acid until energy is available for photosynthesis. In CAM plants stomata open at night so gaseous exchange takes place when it is cool and transpiration is negligible. The carbon dioxide taken in to the plant is stored until it is needed.

If conditions become extremely severe CAM eliminates the need to open stomata even at night. Carbon dioxide released from respiration can be stored and recycled. CAM provides an efficient survival mechanism but it is not very productive.

(b) C_4 Photosynthesis. This is a variation from the normal photosynthetic metabolism which requires a high concentration of carbon dioxide in order to function. In normal photosynthesis (C_3 photosynthesis) the carbon dioxide is channelled directly into the chloroplasts where it is attached to ribulose diphosphate and is converted to a three-carbon acid (see Fig. 43). In the C_4 method, the absorption of carbon dioxide and photosynthesis may take place in separate locations. Concentrations of carbon dioxide are built up by attachment to this P-enolpyruvate carboxylase which is converted to a four-carbon acid. This allows photosynthesis to function when the stomata are only partially open, thus reducing transpiration.

7. Desert heterotrophs. Arid zone ecosystems contain a great range of habitats. Each requires specialisations of lifecycle, morphology and physiology.

(a) Soil. The soil acts as an insulating layer against extreme temperatures. Desert soils have a relatively rich fauna especially near oases. Protozoans can encyst, becoming active for the short time when water is available. Worms and woodlice are limited to damper areas. Molluscs aestivate in deep cracks and are capable of storing sufficient water in their bodies to last for several years.

(b) Temporary pools. Life in desert pools can be very diverse as there are no predators. Many species of insect which require water for their larval stages pass the dry season as drought-resistant eggs. These hatch when water is available and the lifecycle is completed extremely quickly, in some cases within a few days.

Frogs and toads, such as the spadefoot toad of south-western United States, bury themselves deep in the mud of temporary pools before the water evaporates. Between rains they survive on water stored in their bladders, which can form 30 per cent of the body weight.

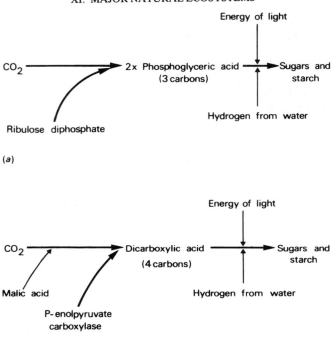

FIG. 43 *Diagrammatic outlines of C_4 and C_3 photosynthetic pathways. (a) Normal C_3 photosynthesis; (b) arid zone C_4 photosynthesis.*

(c) *Rocks and crevices.*

(i) *Arthropods.* Adult insects are scarce between rains but other groups of arthropods such as the scorpions and spiders are prolific. These animals have great powers of water conservation, due both to their physiology and also to their behaviour. Most are nocturnal, spending the days beneath rocks. Urea is excreted as insoluble uric acid and cuticles prevent evaporation from the body surface.

(ii) *Reptiles.* Snakes and lizards are particularly successful in deserts as they tolerate extremely high temperatures. The majority are nocturnal and predatory. Their thick, horny skins contain no glands so water loss is minimal. Many desert lizards have salt glands opening into the nasal cavity, an adaptation which allows them to excrete excess salt taken in with the food in saline areas.

(d) *Burrows.* Small mammals, such as the Arabian jerboa, can survive in deserts by being nocturnal and spending the days in burrows. These generate microclimates of lower temperature and higher humidity (*see* Table XI). Additional adaptations include senses geared for nocturnal life, lack of sweat glands, salt-secreting glands, aestivation and large ears to radiate heat.

TABLE XI. TYPICAL READINGS INSIDE AND OUTSIDE A JERBOA BURROW

	Relative humidity (%)	*Temperature*
Noon		
Surface	5	34.5
Burrow	40	27
Midnight		
Surface	45	23
Burrow	60	22.5

(e) *Exposed areas.* Arid zones contain very few large animals which are forced to live in the open. Antelopes, wild asses, the Arabian oryx and hyenas all have low water requirements and are capable of travelling long distances in search of water.

The camel provides the best example of adaptation to aridity. The hump acts as a food store of fat allowing the rest of the body to function as a radiator for heat loss. Very little water is lost with the excretion of faeces and urea. Camels can tolerate more dehydration than most animals, being able to stand about 30 per cent water loss before the body is strained.

TUNDRA

8. Location. Tundra is a Finnish word meaning open forestless country and is used to describe all types of vegetation in treeless high latitudes. It occupies the zone between 57° latitude (approximately the limit of tree growth) and the polar regions where there is no growing season. Tundra is most extensive in the northern hemisphere due to the lack of land round the south pole (*see* Fig. 44).

9. Environmental conditions.

(*a*) *Temperatures.* At least 7 months of the year experience temperatures below freezing. The average temperature of the coldest month varies from $-10°C$ in the south to $-35°C$ in the north. Frost may occur all year. The growing season lasts for only 2–3 months, and the average temperature of the warmest month is below $10°C$.

(*b*) *Precipitation.* Annual amounts of precipitation are low, mostly between 300 and 500 mm with variations according to the latitude and distance from coasts. Most falls as snow and is extremely effective because of the low rates of evaporation and slow runoff. Most of the tundra is permanently waterlogged.

(*c*) *Day lengths.* In high latitudes day lengths vary greatly through the year. About two-thirds of the tundra zone has continuous daylight in the summer and continuous darkness in the winter. Insolation rates are low but are compensated for partly by the long days in the growing season.

(*d*) *Wind velocity.* Lack of physical barriers to decrease wind velocities results in strong winds which cause damage to vegetation.

(*e*) *Permafrost.* Permanently frozen subsoil is characteristic of the tundra. It forms an impenetrable layer which restricts root growth.

(*f*) *Soil disturbance.* The action of freezing and thawing in the soil causes disruption of the surface layers; this impairs root growth.

10. Ecosystem function in the tundra.

(*a*) *Productivity.* The main constraint on productivity is the low energy input. Primary productivity rates approximate to those of arid zones, that is about 0.5 $g/m^2/year$. Tundra plants have high rates of productivity in the growing season. Most of the plants have high calorie food stores, such as fats, so the energy value of the standing crop is high for its bulk.

(*b*) *Food chains.* The low primary productivity limits secondary productivity. Food chains are short and incorporate few trophic levels. Niches are wide as most species are generalists. The omnivorous habits of the heterotrophs produce complex food webs, and these are accentuated by the fact that many animals

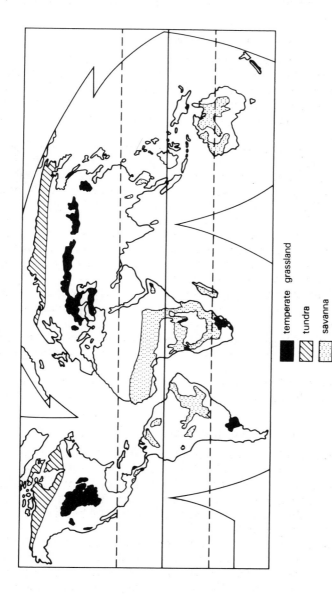

FIG. 44   *The world distribution of tundra and grasslands.*

temperate grassland

tundra

savanna

wander over large areas in search of food. Spasmodic variations in primary productivity are reflected quickly at all trophic levels. Animal populations suffer violent oscillations in numbers because of the instability of the system.

(c) *Nutrient cycling.* The tundra is deficient in nutrients especially nitrates. Decomposition takes place slowly because of the low temperatures and water saturation. Leaf litter may take as long as 3 years to be broken down to humus. Most organisms are longlived so nutrients are held in the standing crop for long periods. Consequently nutrient cycling is slow and impoverished.

(d) *Water stresses.* The functioning of the system is inhibited by water stresses. In winter the ground is frozen so water is unavailable. During spring and early summer physiological drought limits growth. Surface melting is accompanied by high evaporation rates but water absorption by the roots cannot keep pace with the transpiration.

11. Evolution of the tundra climate. The tundra is the youngest of the major world ecosystems, and formed during the Pleistocene glaciations. The tundra was repeatedly split and rejoined during successive glaciations as ecosystems migrated north and south in response to climatic change. The present area of tundra has developed in the 3000 to 10,000 years since the last ice-sheets retreated from the area. Many of its species have circumpolar distributions giving it homogeneity. Tundra organisms have had little time for evolution in this habitat. Most were preadapted in mountain areas or had wide tolerance ranges.

12. Tundra autotrophs.

(a) *General characteristics of the vegetation.* Tundra communities are low growing and lack substantial layering. Heaths, grasses, sedges, mosses and lichens predominate. The proportions of various plant types reflect local differences of climate and drainage. Overall species diversity is low, most having wide distributions.

To the south, the tundra grades into the coniferous forest of the boreal zone. To the north, vegetation cover becomes progressively sparser grading into *feldfields,* where at least 50 per cent of the ground surface is exposed, then ultimately into barren wastes.

(b) *Adaptations.*

(i) *Morphology.* Mat, cushion and trailing growth forms

predominate. These present least resistance to wind. Growing buds are sheltered by the tightly packed stems.

(*ii*) *Physiology.* Many tundra plants have cold resistance. The density of the cell sap increases to resist freezing.

(*iii*) *Reproduction.* Annuals are scarce because of the short growing season. Most species spread reproduction over two or more seasons in order to have sufficient net productivity to form seeds. In the most severe habitats plants reproduce by vegetative means.

Very few plants rely on insect pollination. Most are wind pollinated or self-fertilising. Many species exhibit *vivipary*, that is the seeds start to germinate before they are shed from the parent plant. This helps to ensure seedling survival.

13. Tundra heterotrophs. The permafrost, harsh winters and low primary productivity limit the permanent fauna. Many birds and large mammals, such as the caribou and musk-ox, are migratory, using the tundra as a summer feeding and breeding area.

(*a*) *Soil fauna.* Tundra earthworms of the family Enchytraeidae are adapted for boring through frozen ground. They feed on algae and are relatively abundant. Other soil animals include snails, springtails and insect larvae.

(*b*) *Arthropods.* Insects and spiders are prolific. Most have aquatic larvae and need two summers to complete their breeding cycle. Usually they hibernate over winter as resistant pupae.

(*c*) *Birds, and mammals.* Few birds spend all year in the tundra. Those which do, such as the ptarmigan and grouse, have large feet which act as snow shoes. Few mammal species hibernate, mainly because the growing season is too short to allow the accumulation of fat stores. Mammals at all trophic levels feed through the winter. Small mammals like voles, shrews and lemmings can live beneath the snow. Predators include stoats, arctic foxes, wolves and polar bears. Most of these are euryphagic, even eating vegetation if food is sparse.

14. Instability of the tundra. The tundra ecosystem is disrupted easily by changes in energy input or by interference by humans. This instability derives mainly from the lack of productivity and lack of diversity of the system. Ecologists have asked several questions about this.

(*a*) *Is the diversity limited by the harsh climate?* The wide tolerances and adaptations necessary for life in the tundra may preclude many organisms. However, it is not clear why if some can adapt and survive, others cannot.

(b) *Is the diversity limited by the low energy input?* Lack of productivity limits population sizes and hence the potential for speciation. Areas with few species have few ecological niches.

(c) *Is the diversity limited because of time?* The tundra ecosystem is youthful. The system may not have developed to its full potential within the constraints of energy flow.

GRASSLANDS

15. Location. Grassland ecosystems occupy large areas of the world both in tropical and temperate climates (*see* Fig. 44). Temperate grasslands, known as *prairies* in the new world and *steppes* in the old world, do not have woody plants present. Tropical grasslands, known as *savanna*, usually have some trees in the vegetation.

16. Characteristics of grasses. The grass family, Gramineae, is a highly successful group, having a worldwide distribution. Grasses can withstand grazing and burning because they grow from the base of the leaf, not the apex. The extensive root systems absorb nutrients and water efficiently and stabilise soil. Grasses have a great reproductive capacity since they produce abundant seed which is dispersed widely. Once a grass sward is formed it tends to prevent other plants invading the community.

17. Environmental conditions. Although grasslands occur in a wide range of latitudes they have three environmental features in common.

(a) *Precipitation.* Natural grasslands occur in sub-humid or semi-arid regions characterised by low, variable rainfall. Most precipitation falls in spring and early summer, when potential evapotranspiration rates are high. This combined with a high rate of runoff reduces the effectiveness of rain.

(b) *Topography.* Grassland formations are associated with extensive areas of low surface relief.

(c) *Microclimates.* The vegetation is low growing so there is little obvious structural layering. Climatic amelioration occurs mostly near the soil surface.

18. Ecosystem function in grasslands.

(*a*) *Productivity*. The primary productivity of grasslands is low compared with forests in similar climates. The standing crop is small and has a large percentage of its biomass in the soil. Productivity shows marked seasonality, increasing during the wet season. Savanna areas experience summer droughts when primary productivity is negligible.

(*b*) *Food chains*. As in other areas of restricted productivity, energy flow through the system is low. Food chains are short but complex, incorporating many euryphagic species. A large percentage of the total energy flow passes through food chains in the soil, reflecting the distribution of the biomass.

(*c*) *Nutrient cycling*. Grasses do not hold organic matter in the standing crop for long. Decomposition occurs rapidly so nutrients are circulated through the system quickly. Most grasses are not nutrient demanding. Grasses contain small amounts of potassium, magnesium, phosphorus and nitrogen per gram dry weight compared with forest vegetation. The rate of nutrient turnover is high, therefore, but amounts circulated are relatively low.

19. Temperate grassland autotrophs. Most grasses of the steppes and prairies are perennial and have rolled or folded leaves. Throughout the grasslands the most frequent genera are stipa (needle grasses) and grama (bouteloua grasses). The community composition varies with climate, particularly with rainfall. A threefold division is made on the basis of the height and luxuriance of the vegetation. In North America changes occur with decrease in rainfall from east to west.

(*a*) *True prairie*. This grows to between 2 and 3 m tall. It forms a continuous sward dominated by tussock grasses and has associated herbs such as goldenrod and sunflower. Most of this community has been removed for agriculture.

(*b*) *Mixed prairie*. This occurs west of about 100° longitude. Communities consist of mid-height grasses which grow to approximately 1 m tall and dwarf grasses which grow to only a few centimetres tall.

(*c*) *Short-grass prairie*. This grows in the very arid western states. Grasses are all dwarf and xerophytic. Overgrazing opens the community to invasion by desert shrubs.

20. Savana autotrophs. Savanna grasses are mostly tussocky and perennial. They have flat leaves, are coarse and fast growing.

Trees and shrubs associated with the savanna are fire-resistant, characteristically having thick, corky bark. Herbaceous plants in the communities tend to be xerophytic and to have underground storage organs. Savanna can be divided into four types.

(a) *High grass–low tree*. This is the most luxuriant community and occurs extensively only in Africa where it flanks tropical rain-forests. Grasses, such as elephant grass (*Pennisetum* species), attain heights of over 2 m. Scattered deciduous trees grow to about 10 m tall.

(b) *Tall grass–acacia*. This also has local names, for example *campos* and *laanos*. Communities contain a large variety of tussock grasses growing to about $1\frac{1}{2}$ m tall. Deciduous acacia trees are the most frequent woody plants, except in Australia where eucalyptus predominates.

(c) *Discontinuous xerophilous grass*. This is the most arid type, and contains scattered thorny shrubs and has much bare ground. This type occurs on the edges of deserts.

(d) *Savanna woodland*. This occurs where there has been little human disturbance. Many workers think that it is the climax vegetation type for the climate as it is the most complex, diverse and stable of the savanna communities.

21. Grassland heterotrophs. The most important grazers are mammals. Grasslands present habitats exposed both to predators and to climatic variations. Small mammals such as mice, voles and gophers live in burrows, coming to the surface only for feeding. Large mammals, such as the bison, zebra and antelope, have flocking habits for protection. Many of the grassland herbivores have strong incisors and roughened molars as an adaptation for chewing the tough grass.

A large percentage of grassland animals have migratory habits, reflecting the seasonality of primary productivity in the system.

22. The origins and status of grasslands.

(a) *Climatic climax*. In the early years of this century grasslands were considered to be the climatic climax vegetation in areas too dry for forest. However, attempts to define grassland climates have been abortive, and evidence suggests that grasslands are not solely due to climate.

(i) In many cases there are sharp boundaries between forests and grasslands. If the communities were climatically determined they would grade into one another.

(*ii*) Trees may persist into arid areas yet are absent from some places with sufficient rainfall to support them. Various alternative theories for the origin of grasslands have been suggested.

(*b*) *Fire action.* Woodland could degenerate to grasslands interpersed with fire-resistant trees if the ecosystem were subjected to frequent fires. It is thought that this has been important in the origin of savannas.

(*i*) Fire is an important environmental factor now in savannas, where leaf litter accumulates during the dry season. Trees present are fire-resistant species.

(*ii*) Archaeological evidence indicates that humans have used fires in catching game and in agriculture in savanna areas for over 10,000 years.

(*iii*) If areas of savanna are protected from fire, the percentage of woody plants in the community increases markedly.

(*c*) *Edaphic constraints.*

(*i*) Frequent fires affect the soil, especially the earthworm and microbial populations. Decrease in the soil fauna leads to changes in the nutrient cycling and a decrease in fertility.

(*ii*) In savanna areas silica tends to be leached from the soil whereas aluminium and iron remain. This leads to the formation of *lateritic crusts* on the soil which inhibit plant growth.

(*iii*) Grassland areas are associated with low relief. Little surface erosion takes place on plateaux so leached nutrients are not replenished.

(*d*) *Grazing.* In moderation grazing may increase the proportion of woody growth by reducing competition from grass. If grazing is intensive trees cannot regenerate. Only thorny species survive. Natural grasslands have well-developed grazing fauna which suggests that this is an important aspect of the system.

(*e*) *Climatic change.* Some ecologists interpret grasslands as relicts of a drier climatic regime of either the Tertiary or Quaternary periods. These relict ecosystems may have persisted because of the constraints of fire and soil conditions, accentuated by human activity.

(*f*) *Current opinion.* It is now clear that grassland ecosystems are the result of many interacting factors. The relative importance of the factors has varied in time and place. Although the ecosystems have some relationship with climate they cannot be considered to be climatic climaxes.

FORESTS

23. Features of forests. Forests form the natural dominant vegetation over about two-thirds of the world's land surface. Trees have a great variety of ecological tolerances and inhabit many types of climate. They attain dominance by their size and longevity. A tree canopy determines the microclimate of understorey vegetation and establishes the pattern of nutrient cycling. These conditions are imposed throughout the lifespan of other organisms since trees outlive other life-forms.

Forests are complex ecosystems with much potential for stratification. Most have high rates of productivity and large amounts of biomass in the standing crop. Forest formations exhibit a broad correlation with climatic zones (*see* Fig. 45).

(*a*) *Boreal forest.* This is also known as northern coniferous forest or *taiga,* and occupies the zone from the edge of the tundra to about 800 km southwards.

(*b*) *Temperate deciduous forest.* This occupies areas of temperate climate in mid-latitudes. Its natural distribution extended over most of Europe, eastern North America, east Asia and parts of South America and Australia. Much has been removed by civilisation.

(*c*) *Tropical rain-forest.* This occupies regions of low altitude near the equator.

24. Boreal forest.

(*a*) *Environmental conditions.* Boreal forests grow in regions of cold or cool, moist climates of continental interiors.

(*i*) *Precipitation* is between 375 and 500 mm per annum, mostly falling as snow. Potential evapotranspiration is low so the precipitation is very effective.

(*ii*) *Temperatures* are higher than in the tundra. The mean temperature of the warmest month is above 10°C. The growing season extends for 3 or 4 months. Although insolation is low, day lengths are long in summer because of the high latitude. Severe frosts occur in winter.

(*iii*) *Wind velocity* is decreased by the presence of trees. Beneath the canopy relative humidity remains high so physiological drought is not severe.

25. Ecosystem function in boreal forests.

(*a*) *Productivity* is low (about 3000 kcal/m^2/year) compared

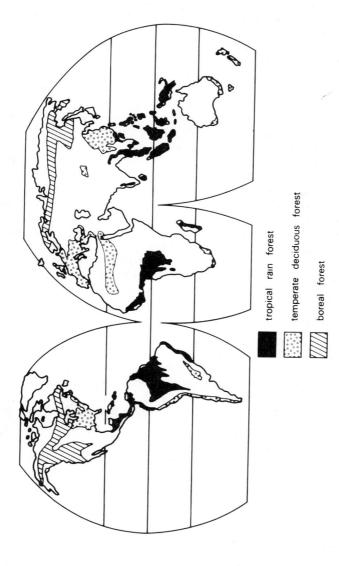

tropical rain forest

temperate deciduous forest

boreal forest

FIG. 45   *The world distribution of major forest types.*

with other forests because of the short growing seasons and low energy input. The continuous vegetation cover results in a fairly high productivity rate for the type of climate because coniferous forests form an efficient photosynthetic surface. This is due to their dense cover, the conical shape of the trees which avoid mutual shading, and their dark colour which absorbs energy.

(b) *Food chains* are short and have few trophic levels. The fauna lacks diversity and has a small biomass because of the limited energy flow. The marked seasonality of primary productivity causes severe oscillations in animal populations.

(c) *Nutrient cycling* is slow and impoverished. Coniferous trees are not usually nutrient-demanding. Leaf litter has a low nutrient content. Decomposition in the cool moist climate is achieved mainly by fungal action. This proceeds slowly and results in mor humus. The large and longlived standing crop locks nutrients in organic matter for long periods. However, the coniferous trees shed leaves continually so there is a steady return of nutrients through the system.

(d) *Soils.* Boreal forests are associated with podsols, with well-developed horizons. The nutrient deficiency, acid conditions and downward movement of water through the soil promote leaching of the surface layers. Redeposition may form hard pans. Leaf litter accumulates on the surface. The soil fauna may contain many small organisms but has few large ones such as earthworms, spiders and snails.

26. Boreal forest autotrophs.

(a) *Communities are homogeneous and lack diversity.* The forest contains few types of dominant tree, most of which have wide distributions. For example, species of pine, spruce and fir are found throughout. These form a dense canopy which limits the development of undergrowth. Understorey layers are sparse. Associated shrubs include laurel, dogwood and willow. In addition a few herbaceous plants, such as dwarf cornel and buttercup, may be present.

The boreal forest becomes progressively more stunted northwards until it gives way to birch and larch which form a transitionary ecosystem (ecotone) next to the tundra.

(b) *Adaptations of plants to endure the severe winter and short growing season.* These include the following.

(i) Increasing the concentration of cell sap to reduce its freezing point.

(*ii*) The evergreen habit which ensures that leaves are ready to function as soon as temperatures permit.

(*iii*) Needle leaves which may protect against cold and desiccation.

(*iv*) Flexible structure of trees which do not break under the weight of snow.

27. Boreal forest heterotrophs. Very few animals rely exclusively on the boreal forest for food. Invertebrates are very important both in the soil and in the tree canopy. Many insects, for example the spruce budworm and the eastern spruce beetle, are destructive. Large stands of single species of trees present ideal habitats for them. The seeds of conifers provide food for small mammals like squirrels and birds such as the crossbills.

Most larger animals, including the moose, caribou and snowshoe rabbit, tend to feed in seral areas where foliage is more luxuriant. Carnivores like the wolf and lynx often feed on small mammals and travel long distances in search of prey. Many secondary consumers are omnivorous which is an ecological advantage when food is sparse.

28. Temperate deciduous forest.

(*a*) *Environmental conditions.* Temperate deciduous forest occupies areas without extremes of temperature but still with marked seasons. Rainfall is moderate, between 760 and 1500 mm per annum. The growing season extends over about six months and insolation is higher than in boreal areas. Most areas of temperate deciduous forest have been modified by human activity and do not represent truly natural ecosystems.

29. Ecosystem function in temperate deciduous forest.

(*a*) *Productivity* is higher than in coniferous forest but less than in tropical forests. Primary productivity is about 8000 kcal/m^2/year.

(*b*) *Food chains* have many trophic levels because primary productivity is high. Food webs are complex, involving many specialist feeders. The detrital food path is usually more important than the grazing food chain.

(*c*) *Nutrient cycling* varies with the species of tree present. Most deciduous trees are nutrient-demanding and produce nutrient-rich litter. The climate allows bacterial decomposition which is rapid and produces mull humus. The deciduous habit returns most nutrients to the soil in the autumn making the cycling of nutrients spasmodic.

(*d*) *Soils* are mostly nutrient rich with well-developed horizons. Characteristically, temperate forests grow on brown earths or rendzinas. These have negligible leaching and are either neutral or alkaline. The soil fauna is prolific.

30. Autotrophs. *Vegetation communities* are far more diverse than in boreal forest. Areas of forest are isolated and so exhibit differences in the dominant species. The European forests have approximately twelve dominant species including oaks, beech, ash and chestnut. In contrast North American forests are richer having about 60 dominant tree types. The distribution of the dominants varies with local conditions.

In all cases the forests are associated with prolific undergrowth and the vegetation structure may have several layers. The composition of these depends on the amount of energy penetrating the canopy. Shrubs such as willow, hazel and hawthorn form discontinuous understorey layers, while a wide variety of herbs such as wood anemones, violets and bluebells form the ground layer.

Adaptations to the ecosystem type are found both in the dominant and associated plants. All species must survive the cold of winter. Plants adopt the strategies of overwintering described in V. Many understorey plants complete their lifecycles early in the growing season before the leaf canopy is formed on the trees and intercepts the light.

31. Heterotrophs. These are diverse and abundant. The complex physical structure of the forest presents a great variety of habitats including shoots, leaf litter, bark, wood and flowers. The diversity of habitat and large energy flow increases competition between organisms and encourages specialisation of niches. This is evident at all trophic levels. The complexity of the food chains gives the system stability so populations suffer smaller and less frequent oscillations than those in boreal forests do.

32. Tropical rain-forest.

(*a*) *Environmental conditions*. Annual rainfall exceeds 2000 mm falling throughout the year usually with one or more relatively dry periods. Temperatures and insolation rates are high with very little seasonal variation. Relative humidity is consistently high.

(b) *Ecosystem function.*

(i) *Productivity.* The potential for continuous growth in the wet equatorial climate makes tropical rain-forests one of the most productive ecosystem types in the world. Primary productivity approximates to 20,000 kcal/m^2/year and supports a large amount of animal biomass.

(ii) *Food chains* are long and extremely complex. Specialist organisms predominate because of the high rate of energy flow and the intense competition between species.

(iii) *Nutrient cycling* takes place rapidly and involves large amounts of nutrients. Decomposition occurs quickly by bacterial action so there is little storage of litter in the system. The evergreen broad-leafed trees give a constant return of nutrients to the soil.

(iv) *Soils* are fertile in the undisturbed forest system. The heavy precipitation could cause leaching but it is compensated by high rates of evaporation. Once the forest canopy is removed organic matter is oxidised quickly and fertility is lost.

33. Tropical rain-forest autotrophs.

(a) *Vegetation communities.* The main areas of tropical rain-forest are physically isolated and so contain different genera. However they exhibit great similarity in structure and adaptations. All are dominated by broad-leafed evergreen trees and are extremely diverse with many specialist species. The rain-forests are highly stratified. Typically the trees form three layers.

(i) Scattered very tall trees that grow above the general canopy level.

(ii) Canopy layer of continuous growth about 30 m tall.

(iii) Understorey layer of discontinuous growth.

The combined thickness of these layers intercepts much light so little energy is left for ground layer vegetation. Epiphytes (plants which grow on other plants) and climbing plants abound because these strategies enable plants to reach the light in the canopy layers.

(b) *Adaptations.* In the condition of intense competition few adaptations have been successful. This has led to evolutionary convergence.

(i) The evergreen habit allows maximum primary productivity per annum. Deciduous trees are at a competitive disadvantage.

(ii) Leaves tend to be dark green, possibly to absorb the maximum amount of light. Their leathery textures could protect them against the high temperatures and insolation. Many leaves have extended tips which may help water loss from the surface by facilitating dripping.

(*iii*) *Buttress roots*. Many trees have plank-like projections from the base of the trunk to the soil. Originally it was thought that these increased stability but this is now uncertain.

(*iv*) *Cauliflory*. Flowers and fruit tend to grow directly from the trunks of trees as in cacao, for example. The advantage is obscure; possibly it aids pollination and seed dispersal.

(*v*) *Regeneration*. Tree seedlings are adapted for slow growth as sciophytes until a space occurs in the canopy. When exposed to bright light, the seedlings function as heliophytes and grow rapidly.

34. Tropical rain-forest heterotrophs. A larger proportion of animals lives in the canopy layers of tropical forests than in temperate forests, reflecting the distribution of primary productivity. In addition to arboreal mammals, there are arboreal snakes and amphibians. The rate of speciation has been especially high producing immense numbers of species at all trophic levels. Although some animals are brightly coloured most are inconspicuous or are nocturnal in order to avoid predators. Symbionts and parasites are prolific. Animal numbers do oscillate but not as markedly as in other systems. Provided that the ecosystem is undisturbed it is relatively stable. However, the balance is precarious and can be disrupted easily.

THE MARINE ECOSYSTEM

35. Introduction. Oceans cover about 70 per cent of the world's surface and support a biomass estimated to be ten times greater than that on land. The marine ecosystem contrasts with terrestrial ones in its environmental conditions and structure, but the basic functioning is the same. The study of marine ecology is becoming increasingly important in the context of exploiting the sea's food reserves.

36. Environmental conditions. The aquatic habitat is more favourable for life than is land. There is no danger of desiccation and essential nutrients, oxygen and carbon dioxide are readily available dissolved in the water. As far as is known, no abiotic areas occur in the sea. It is a continuous medium but contains definite ecological barriers such as temperature, light penetration, pressure and salinity.

Salinity determines the distribution of many species. It is caused by at least 45 elements, including sodium, chlorine, magnesium and bromine, and varies with depth and location.

Extremes of temperature are rare in the sea. Temperatures are uniform over extensive areas because heat is redistributed by vertical and horizontal currents.

37. Ecosystem function in the seas.

(a) *Productivity*. The main limits to primary productivity are imposed by light intensity and nutrient supply. The extent of light penetration varies with turbulence, the angle of the rays, and the amount of suspended matter in the water. The depth at which light is sufficient for photosynthesis is the *euphotic* zone.

In inshore waters it may be as shallow as 5 m but at maximum is about 100 m which corresponds to the depth of continental shelves (*see* Fig. 46). Deeper than this, the *disphotic* zone, which is lit dimly, extends to about 200 m deep. Below this the seas are completely dark.

Mean annual primary productivity is estimated to be between 25 and 50 per cent less than that on land. This is mainly because of the deficiency of nitrogen, phosphorus and potassium in the euphotic zone, and because of the lower net productivity of marine plants.

(b) *Food chains and marine organisms*. The marine flora is less diverse than that on land. Algae, of which phytoplankton are the most important, predominate. These unicellular plants float near the surface of the sea. They are highly productive but have short lifecycles so the biomass at any one time may be low. For this reason animal biomass usually exceeds plant biomass.

Food chains tend to be long and complex because the marine fauna is very diverse. The grazing food chain is more important than the detrital food chain.

Most of the phytoplankton are consumed by zooplankton. Large herbivores are rare because of the microscopic size of the plants. The zooplankton form the major link between primary productivity and other forms of marine life. They are consumed by surface water fish, such as mackerel and herring, and form the main diet of baleen whales. In addition the zooplankton are eaten by bottom-living invertebrates which are, in turn, eaten by bottom-living fish.

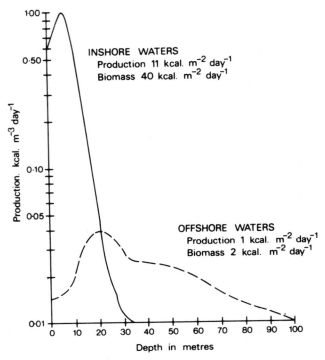

FIG. 46   *The vertical distribution of primary production in the sea.*

(c) *Nutrient cycling.* Productivity is frequently limited by nutrient supply because elements are removed rapidly by the phytoplankton in the surface layers. Nutrients may pass through surface food chains but are ultimately lost to the ocean floor where they are liberated from organic matter by decomposers. Primary productivity is greatest where upwelling currents return the nutrients to the surface layers.

Productivity and nutrient cycling show marked seasonality in high latitudes. In winter nutrients accumulate because growth is negligible. The ocean layers become mixed by turbulence caused by storms. A peak of production occurs in spring (*see* Fig. 47) as nutrients are available.

In summer the oceans develop vertical stratification of temperature. A sharp junction, known as the *thermocline*, occurs between warm surface layers and cool bottom layers. Little

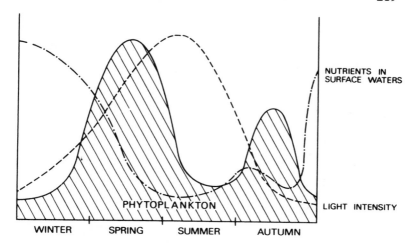

NUTRIENTS IN
SURFACE WATERS

LIGHT INTENSITY

PHYTOPLANKTON

WINTER   SPRING   SUMMER   AUTUMN

FIG. 47 *The pattern of primary productivity in relation to light intensity and nutrient supply in surface waters of high latitudes in the northern hemisphere.*

mixing takes place across this barrier so nutrients are not recycled.

In autumn the surface layers cool and sink, causing mixing and the recycling of nutrients. This results in a second peak in productivity. In tropical areas growth continues all year but the thermocline may be permanent so nutrient recycling is slow.

PROGRESS TEST 11

1. What are the basic features of desert ecosystems? **(2)**
2. Why are arid zone food webs complex? **(4)**
3. What is crassulacean acid metabolism? **(6)**
4. Why are reptiles successful in deserts? **(7)**
5. What are the main features of the tundra habitat? **(9)**
6. Why is nutrient cycling slow in the tundra? **(10)**
7. Why is the tundra ecosystem instable? **(14)**
8. What is the difference between steppes (or prairies) and savannas? **(15)**
9. What were the original three main types of prairie in North America? **(19)**
10. Why do ecologists now think that natural grasslands are not climatic climaxes? **(22)**

11. What are the three main types of forest ecosystem?   **(23)**

12. Why is productivity relatively low in boreal forests?   **(25)**

13. Why are heterotrophs sparse in boreal forests?   **(25,27)**

14. How does nutrient cycling in temperate deciduous forests differ from that in boreal forests?   **(25,29)**

15. Why are food chains long in tropical rain-forests?   **(32)**

16. Give four examples of common adaptations of plants into the tropical rain-forest environment.   **(33)**

17. What is the difference between the euphotic zone and the disphotic zone in the sea?   **(37)**

18. How does nutrient cycling limit primary productivity in marine ecosystems?   **(37)**

CHAPTER XII

Selected British Habitats

HISTORICAL DEVELOPMENT

1. The postglacial history of British vegetation. The present flora and fauna of Britain are mainly the result of migrations and land-use since the last glaciation. Ice-sheets last retreated from Britain approximately 14,000 years ago. Since then the climate has warmed progressively and has undergone a series of changes (*see* Table XII).

Plants and animals migrated back into Britain across land bridges from the continent as the climate improved. The sequence of species returning reflected the conditions prevailing and the time since glaciation. Each climatic period was characterised by associated vegetation communities. These can be determined from deposits of pollen in sediments.

2. The sequence of communities. The arctic wastes and tundra of the late glacial developed into pine and hazel woodlands by the boreal period. These gave way to mixed forest in the Atlantic period (*see* Table XII). The land bridge to the continent was submerged by the rising sea level in about 5500 BC; after this, further immigration was more difficult.

The Atlantic forest formation represented the climax vegetation of Britain. After this time vegetation became increasingly modified by human activity, and very few remnants of the forest formation remain. Most ecosytems in Britain are either "man-made" or "semi-natural".

Overall the flora and fauna of Britain are poor both in species and variety in comparison with that of Europe. This is mainly the result of disruptions caused by glaciations and the problems involved in recolonisation.

WOODLANDS

3. Features of woodlands. At present woodlands cover only about 1 per cent of the area of Britain but they are more prolific in

TABLE XII. GENERALISED SEQUENCE OF LATE-GLACIAL AND POSTGLACIAL CHANGES OF VEGETATION AND CLIMATE IN BRITAIN

Dating	Periods	Pollen zone	Vegetation	Forest cover	Climate	Cultures
AD 2000	Sub-atlantic	VIII	decline of forest. Increase semi-natural ecosystems		cold and wet more oceanic	Norman Anglo-saxon Romano-British Iron age
BC 500	Sub-boreal	VII	mixed forest maximum		warm and dry, continental	Bronze age Neolithic
Postglacial 3000				FOREST	climatic optimum, warm and wet	
5000	Atlantic	VI	alder, oak elm, lime			
7500	Boreal	V	oak, elm, pine hazel		warm and dry	Mesolithic
8000	Preboreal	IV	pine, birch hazel, birch, pine birch scrub		less cold sub-arctic	
8300	Younger dryas	III	birch tundra heath		tundra	
Late glacial	Allerod	II			milder	
13000	Older dryas	I	open tundra arctic waste		sub-arctic	Palaeolithic

certain areas. For example, they cover nearly 8 per cent of south-eastern England. Most woodlands in Britain are either planted or are derived from planted woodlands. Practically all of these semi-natural ecosystems have undergone some management for timber production or for hunting. The functioning of forest ecosystems is described in XI. Attention is given here to the types of woodlands in Britain and to the management practices which have been important in their formation.

4. Composition of deciduous woods. These represent part of the European temperate forest formation. The main dominants of British deciduous woodlands are oak, beech, ash and sycamore with holly, willow, hazel and hawthorn as the principal understorey shrubs. Typically these woodlands contain four layers or strata (*see* Fig. 48). The two species of oak occurring naturally in Britain, *Quercus petraea* (the sessile oak) and *Quercus robur* (the pedunculate oak) are the most important dominants and can grow in a variety of site conditions. The accompanying species vary more with differences in soil, nutrients and drainage.

Most British woodlands are dominated by one or two species of tree. In many cases this is a legacy of either planting policy or selective felling in the past.

5. Regeneration of deciduous woods. Deciduous woodlands in Britain tend to have even aged canopies and lack natural regeneration. This is due partly to pressure from human interference and partly to the nature of the plants themselves.

(*a*) The dominant trees are slow to mature. Oak and beech may be 40 or 50 years old before they flower.

(*b*) Seed production of the dominants is variable and spasmodic.

(*c*) It is difficult for seedlings to become established and to survive under the canopies because of shading, root competition and damage from herbivores.

6. Ancient woodlands. Some woodlands in Britain are remnants of royal forests established by the Normans for hunting. In the twelfth century, over one-third of the country was covered by royal forests. Most have been removed for agriculture but some areas, such as the New Forest and Sherwood Forest, have been relatively undisturbed. These contain ancient trees and many rare species of plants and animals.

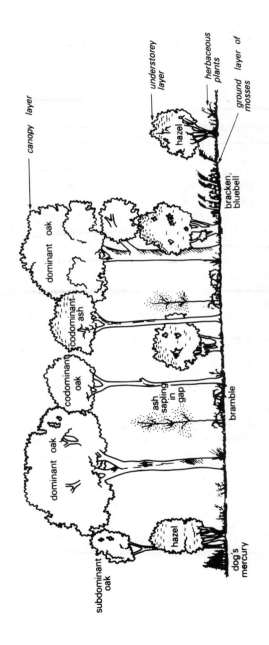

FIG. 48 *Profile of an oak wood.*

7. Managed woodlands. Many deciduous woodlands have been managed for timber production. Although this activity tends to be uneconomic today, evidence of it can be seen in the structure of the woodlands. Two practices were most important. In both of them usable timber is obtained by suppression of the terminal bud.

(*a*) *Coppicing.* This began sometime in the middle ages and continued as an economic practice until after the second world war. The method produces tall, thin, straight poles for fences, firewood, hurdles and charcoal. Hazel or sweet chestnut is cut to ground level. Regrowth is rapid and produces straight branches (*see* Fig. 49). Traditionally, the coppiced wood is grown beneath a canopy of standard trees, either oak or ash, spaced at a density of 6 per hectare (twelve per acre) so that their crowns do not overlap. The standards produce wide boughs used originally for ships' timbers. The coppice is cut every 12–15 years. Many coppices have rich communities of herbaceous plants.

(*b*) *Pollarding.* Several deciduous trees such as beech and oak regenerate their canopies if the branches are cut off at the top of the trunk. This regrowth produces a dense crown of small timber at a height safe from browsing animals. Pollarding was carried out extensively from the middle ages to about the mid-nineteenth century.

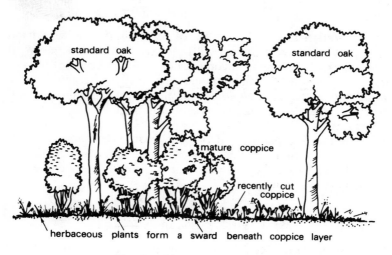

FIG. 49 *Profile of a coppiced wood.*

8. Coniferous woods. Coniferous woodlands in Britain represent an ecotone between the temperate deciduous forest and the boreal forest. They occur mainly on well-drained acid soils. Few areas of ancient coniferous forest remain, as for example at Rothiemurchus in the Spey valley, Scotland.

Conifers regenerate successfully in many places. The dominants mature quicker than those of deciduous woods and have a higher net productivity in the early years of growth. This, together with the lightness of the wood and its good physical qualities for working, make conifers important commercially.

9. The forestry commission

(a) *Formation*. The first world war showed that our woodlands could not sustain us even for a few years if timber imports were cut off. The forestry commission was formed to re-establish forest resources on an adequate scale.

(b) *Plantation policy*. Conifers were planted extensively in areas of marginal agricultural land with low productivity. Sitka spruce and Norway spruce were used frequently on drained bogs, and lodgepole pine was planted in sites at high altitude or with low nutrient status. Most early post-second world war plantations were of single species in large rectangular blocks. Public opinion and enlightened attitudes have caused the commission to change its policies. Most modern plantations have deciduous screens round their edges and are shaped in relation to contours.

10. Plantation habitats. As a lot of plantations were grown on land previously used for agriculture a true woodland flora and fauna took a long time to become established. The plantations are important habitats for many species, especially in Scotland where woodland is sparse. Red squirrels, pine martens, wild cats and crossbills have all increased in numbers as a result of afforestation.

If conifers are planted on old deciduous woodland sites the change in nutrient cycling (*see* XI) induces changes in soil characteristics leading to increased acidity and decreased diversity of soil fauna.

HEATHLANDS

11. Introduction. Heathland vegetation is dominated by evergreen dwarf shrubs of the heather family (Ericaceae). Trees, tall shrubs and extensive grass turf are absent.

Heathlands were very widespread in northern Europe until early this century. They formed a distinct community type often used for common grazing land. Traditionally, heaths are burnt every few years to remove old growth and to encourage new shoots for herbage. The area of heathlands has decreased greatly since the nineteenth century as their use has become uneconomic. If they are not burnt or grazed, succession takes place and they develop into woodlands.

12. Environmental conditions.

(a) *Climate*. Heathlands occur in areas of moist temperate climate with mild winters where at least 4 months have average temperatures above 10°C. The temperature of the warmest month is below 22° C. Relative humidity is fairly high and periods of drought are short.

(b) *Soils*. These are freely drained and acidic. Heathlands are associated with podsols with thin peat layers at the surface.

(c) *Exposure*. Heath plants can tolerate extreme exposure, as for example on cliffs and mountains, but they cannot tolerate shading from taller plants.

(d) *Grazing and burning*. The plants can withstand these pressures because of their morphology and lifecycles (*see* **15(b)**).

13. Status.
In a few cases, such as on exposed cliffs or on mountains above the treeline, heaths are the natural climax vegetation. However, most heathlands are the result of human activity. The development of the ecosystem to its full biotic potential is checked by the action of fire and grazing. The heathlands are unstable systems and undergo rapid change if the arresting factors are removed.

14. Origins.
Many theories have been formulated to account for the origins of heathlands. The importance of human action in maintaining the heaths is undeniable but controversy existed as to how heathlands developed in the first instance. Clearly they are not determined climatically since the climate type supports forests. Similarly the associated soils are capable of supporting trees.

(a) *Seral theory*. In the 1940s many workers, including the British plant geographer Sir Arthur Tansley, thought that the heaths were a seral stage in the post glacial development of vegetation communities. They considered that heathland areas developed directly from tundra without intervening forest. These workers were strongly influenced by the concept of succession.

(b) *Degenerate theory.* As early as 1901, Graebner proposed that the heathlands of northern Germany were degenerate systems derived from forest by progressive felling. The development of pollen analysis in the 1950s produced evidence to confirm this view. Dimbleby (1962) has shown that many of the British heathlands were formed from forests in the Atlantic and sub-boreal periods. The area of heathland continued to expand until the eighteenth century.

15. Heathland autotrophs.

(a) *Vegetation.* Heathland communities are homogeneous and lack diversity in their flora. British heaths have three main dominants, ling (*Calluna vulgaris*), which occurs throughout, bell heather (*Erica cinerea*) which prefers drier areas, and cross-leaved heath (*Erica tetralix*) which prefers wetter areas. These form a canopy beneath which few plants are present in association.

Understorey plants include bilberry (*Vaccinium myrtillus*) and calcifuge grasses such as mat grass (*Nardus stricta*) and wavy hair grass (*Deschampsia flexuosa*) as well as broad-leafed herbs, such as sheep's sorrel (*Rumex acetosella*) and heath bedstraw (*Galium hiercynium*). Mosses and lichens are prolific, growing directly on the peat surface.

(b) *Adaptations to the habitat.* Ling possesses a number of features which make it very successful in heathland.

(i) *Germination.* Seeds are produced in great numbers and have a high percentage viability. They have staggered dispersal and germination, that is development is delayed in some seeds so the potential for regeneration is conserved. Germination can take place over a wide variety of conditions.

(ii) *Burn resistance.* Exposure to moderate heat, as from a local burning, accelerates germination of seeds. Adult plants can regrow from the roots if aerial parts are burnt off.

(iii) *Desiccation resistance.* Adult plants have xeromorphic leaves. Stomata are located in a groove on the underside of the leaf which can be opened or closed to control transpiration. Ling is adapted physiologically to tolerate a low water content in its tissues.

(iv) *Calcifuge habit.* Ling achieves vigorous growth in soils of pH 3.5–6.5. This is mainly due to its association with an endotrophic mycorrhizal fungus, phoma, which is limited to this acidity range.

16. Heathland heterotrophs. The low growing homogeneous vegetation provides little variety of microclimates and habitat potential. However, since heathlands are on soils which warm up rapidly, they have rich invertebrate, reptile and bird fauna.

(*a*) *Invertebrates.* Only a few groups are really scarce on heathlands. These are: molluscs, which are limited to calcareous areas for shell growth; earthworms, which cannot tolerate high acidity; and woodlice, which cannot tolerate very dry conditions. Most other groups of invertebrates such as the ants, weevils, spiders, wasps and mites are prolific, and occupy a variety of niches and trophic levels. For example, within the beetle group some (such as the heather beetle) are herbivorous, and some (such as the tiger beetle) are predatory.

(*b*) *Vertebrates.* The main grazers on heathlands are sheep, cattle, ponies and deer. Ecologically the most interesting groups present are the reptiles and the birds. Some of our rarest reptiles, such as the smooth lizard, are confined to the warm soils of the heathlands. The prolific invertebrate fauna provides abundant food for birds. Many rare species, such as the Dartford Warbler, feed and nest in the heathlands. Others such as the buzzard and the hobby feed in the heath but nest in adjacent woodlands. Heaths in upland areas are important for game birds like grouse and ptarmigan.

17. Heathland function.

(*a*) *Productivity and food chains.* Primary productivity is low especially when the heath dominants are old. The standing crop at both the primary and secondary levels is small, reflecting the low energy flow through the system. Food chains are short but complex webs exist, particularly those involving invertebrates.

(*b*) *Nutrient cycling.* This is low and impoverished with many net losses from the system (*see* Fig. 50). Decomposition occurs slowly in the acid conditions as leaf litter decays mostly as a result of fungal action. Raw humus accumulates to form a peaty layer on the soil surface. Grazing and burning extract nutrients from the ecosystem accentuating the nutrient deficiency. Heathlands are particularly lacking in calcium. Continual removal of supplies from the system by grazing without return of nutrients leads to increased acidity.

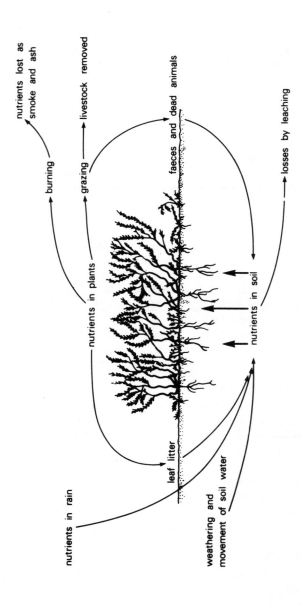

FIG. 50   *The pattern of nutrient cycling in heathland.*

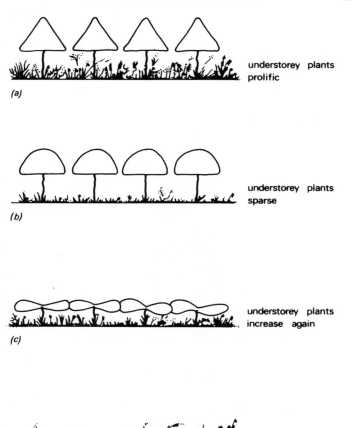

FIG. 51 *Cyclic phases of growth in heathland. (a) Pioneer phase;*
(b) building phase; (c) mature phase; (d) degenerate phase.

(*c*) *Cyclic development.* During the lifespan of ling, changes
occur in net productivity and morphology (*see* Fig. 51).

(*i*) *The pioneer phase* lasts from the seedling stage until the
plant is between 3 and 6 years old. The plant has a pyramid shape
that allows light to penetrate through the canopy. Associated
species are prolific.

(*ii*) *The building phase* lasts until the plant is about 15 years old, and is the period of maximum net productivity. The plant develops a hemispherical shape blocking out light so that few species can exist beneath the canopy.

(*iii*) *The mature phase* lasts until the plant is about 25 years old. Net productivity declines, the plant spreads out from the centre allowing light to penetrate so the diversity of associated species increases again.

(*iv*) *The degenerate phase* lasts until the plant is about 30 years old. During this final stage, the centre of the plant tends to die off, creating a gap which will be colonised by new dominant plants.

Changes in species diversity, food chains and nutrient cyclic parallel these stages of development in the dominant plants.

(*d*) *Management.* Traditional management by burning developed to ensure that the dominant plants were kept at the building phase of growth when they are most productive. Regular burning has ceased on most heathlands in lowland Britain. Some are burnt periodically to halt succession so that the areas are kept open for recreation. Some in nature reserves are subjected to small local burns so that all stages of heath development are present. This encourages diversity of habitat important for conservation. Many heathlands in highland Britain are managed for game, particularly grouse. In these areas regular burnings still take place.

CHALK GRASSLANDS

18. Introduction. British grasslands can be divided into three basic types.

(*a*) *Calcareous or alkaline,* growing on chalk or limestone soils.

(*b*) *Acidic,* growing on the soils of acidic rocks or in highland areas.

(*c*) *Neutral,* growing on soils of moderate pH values and sharing some of the features of both of the other types.

The grasslands vary greatly in their ecological interest. Some have been planted recently while others are of great antiquity and have a long history of management. The calcareous grasslands, especially those on chalk, are the most important for their flora and associated fauna. They contain many of Britain's rarest species.

19. Location. Chalk deposits are widespread in southern England and north-west France. Most develop alkaline soils suitable for calcareous grasslands. In Britain chalk grasslands are found on the North and South Downs, the Chilterns, and Salisbury Plain, together with parts of Wiltshire, Berkshire and Dorset.

20. Environmental conditions.

(a) Soils contain a high proportion of free calcium carbonate and exchangeable calcium giving them high pH values. Most are shallow rendzinas (see VI) with little organic matter in the A horizons. The soils are freely drained and well aerated.

(b) The porosity of chalk and its associated soils leads to water deficiency. The grasslands have little potential for microclimatic amelioration so that the biotic components of the system are prone to desiccation.

(c) Grazing is an important factor in the maintenance of the grasslands. Vegetation must withstand trampling and nibbling especially from sheep, cattle and rabbits.

21. Origins and status. Evidence from pollen analysis and archaeology suggests that the Atlantic forest cover was removed from many chalk areas by 2500 BC. Grasslands were developed and were maintained by human activity. Originally the grasslands were grazed by a variety of domestic stock including oxen and pigs.

A tradition of sheep grazing evolved as wool became an important fibre. If grazing stops the grasslands undergo secondary succession and woodlands develop. This has occurred in many places since the eighteenth century. Woodlands dominated by beech, yew, oak or ash are abundant now in areas formerly occupied by chalk grassland.

22. Chalk grassland autotrophs.

(a) *Vegetation.* Communities are extremely rich and diverse. It is estimated that over half of the species of seed plants and ferns in Britain grow in this habitat. Over 50 species are confined exclusively to calcareous grasslands, including some of the rarest orchids.

The species composition and abundance varies between different areas according to differences in local climate and land-use, but all chalk grasslands have short springy turf. Throughout the main dominant grasses are sheep's fescue (*Festuca ovina*), red fescue (*Festuca rubra*), crested hair grass (*Koeleria cristata*) and quaking grass (*Briza media*). These grow with a great variety of other grasses, sedges, broad-leafed herbs and mosses. However, they all have certain features in common and display similar geographical affinities.

(*b*) *Adaptations to the habitat.* The exacting environment of the chalk grasslands has produced convergence in the features of the plants.

(*i*) Many are *hemicryptophytes*. These have their growing points close to the ground and can sprout again if the shoots are grazed.

(*ii*) *Rosette habit.* Approximately 35 per cent of the chalk sward species grow in a rosette form. This is resistant to grazing and trampling and helps to avoid desiccation.

(*iii*) *Mat habit.* This adaptation is also very frequent. Extensive horizontal branching eliminates competition and aids the plant in surviving grazing.

(*iv*) *Xeromorphism* is important in this dry habitat. Features include narrow or rolling leaves, hairs to reduce transpiration and thick, waxy cuticles.

(*v*) *Storage organs,* such as bulbs, are prolific and act as a buffer against environmental stress.

(*vi*) *Long life.* Most species are perennial. The sward forms a dense mat which precludes invasion or re-establishment by seedlings.

(*c*) *Calcicoles.* Many of the chalk grassland plants are restricted to this habitat because they are calcicoles; these are species limited to alkaline soils. They may be either chemical calcicoles, that is, restricted to this soil type because of its chemical properties, or physical calcicoles, that is, restricted to this soil type because of its physical properties (these include free drainage, warmth and good aeration). Many of the British physical calcicoles are at the limit of their distribution range in this country. Elsewhere they can grow on a variety of soil types but here they are restricted to the most favourable habitat.

23. Relationships with climate.

(*a*) *Geographical affinities.* Several different geographical elements combine to form the flora (*see* Table XIII). Many of the species of the continental elements are adapted to warmer drier climates.

TABLE XIII. GEOGRAPHICAL ELEMENTS IN THE FLORA OF
BRITISH CHALK GRASSLANDS

Approximate percentage of flora(%)	Element	Associated climate	Example species
55	Wide European	temperate	hoary plantain (*Plantago media*)
25	Southern continental		traveller's joy (*Clematis vitalba*)
12	Continental	hot dry summers, cool winters	stemless thistle (*Cirsium acaule*)
4	Northern continental		
3	Southern oceanic	warm, moist	chalk milkwort (*Polygala calcarea*)
1	Northern oceanic	cool, moist	wild thyme (*Thymus drucei*)

(*b*) *Changes in flora in relation to local climate.* The geographical elements display slightly different distribution patterns within the chalk grassland habitats. This can be seen both on regional and local scales. Species with marked southern affinities are most prolific in the south-east. Wide European and northern continental species are most prolific in the northern grasslands. North-facing slopes tend to have far more species than south-facing slopes because the latter become very hot and dry in summer.

24. Chalk grassland heterotrophs. The invertebrate fauna of chalklands is extremely rich and diverse. The varied vegetation communities present many niches. Soils warm up quickly during the day and can support a lot of species which are on the edge of their distribution range in Britain.

Many of the invertebrates are specialists being restricted to a few types of food. For example the large blue butterfly (*Maculinerea arion*) feeds on thyme when young, then on ant larvae. The small blue butterfly (*Anthyllis vulneraria*) feeds on kidney vetch. This specialisation makes them vulnerable if habitats change.

The main grazers are sheep and rabbits. Both of these feed selectively showing distinct preferences for certain plants.

25. Ecosystem function.

(*a*) *Productivity and food chains.* Primary productivity is generally low because of the habitat constraints. The amount of biomass in the standing crop of autotrophs is small, but the variety of components provide many ways for animals to obtain food. Food chains are complex with many intimate relationships and narrow niches. Food webs involving invertebrates can have many trophic levels because little energy is required to support each individual.

(*b*) *Nutrient cycling.* Dead organic matter decays rapidly in this habitat. Most decomposition occurs through bacterial action and produces mull humus in the soil. The majority of plants are perennial so nutrient cycling may be slow. The ecosystem does not suffer great deficiencies of nutrients except in very dry areas when some elements, such as exchangeable potassium, may be lacking.

26. Succession and management. If the grasslands are not grazed secondary succession takes place. Tall grasses and broad-leafed herbs shade out the rosette and mat plants. Mixed scrub composed typically of hawthorn, dogwood and privet develops eliminating all the short growing plants. Eventually woodlands take over as the last seral stage.

All the chalkland species of greatest ecological interest are associated with the short grass phase. Management for conservation aims to halt succession at this level. If grazing cannot be carried out for economic or recreation reasons artificial methods such as mowing, burning or scrub clearance may be used.

SALT MARSHES

27. Introduction. Salt marshes are areas subject to periodic flooding by the sea. They occur on coasts with shallow gradients and in estuaries behind spits. Sediments are deposited on the marsh by the action of tides and drainage from inland. The sediments become established by the salt marsh vegetation so that the deposits accumulate and the marsh grows seaward as the land level rises.

28. Environmental conditions. All habitat factors vary over the marsh in relation to distance from mean sea level.

(*a*) *Tides*. The habitat is flooded at each high tide and is exposed at low tide. This alternation produces environmental stresses for life. For example, flooding reduces the aeration of roots and the photosynthesis of submerged plants, whereas exposure may lead to desiccation.

(*b*) *Salinity*. This varies over the marsh both in space and time. Most importantly it varies with the state of the tide, rainfall, distance from drainage creeks and drainage from inland. The presence of salt inhibits the uptake of nutrients and water.

(*c*) *Soils*. These are generally formed on a sand base. Particle size depends on the conditions for sedimentaion, including distance from creeks. Coarser materials are deposited first as finer materials are carried further. Because of this, banks build up on the edges of creeks. The soil never becomes completely waterlogged. There is always a shallow aerated layer just below the surface even during flooding.

(*d*) *Exposure*. The amount of time that the marsh surface is exposed to air each day varies with distance from the mean sea level and the drainage creeks.

29. Salt marsh autotrophs.

(*a*) *Vegetation communities*. Very few higher plants can tolerate the saline conditions so the flora of salt marshes is very specific. Communities vary with environmental factors. Vegetation tends to be zoned from the low water level inland reflecting the time of exposure and salinity (*see* Fig. 52). Each zonal community can be regarded as a sere in a succession from bare mud or sand to dry land. The species of each sere tolerate the conditions within that tidal range. The communities migrate seaward as the marsh grows.

The broad zonal pattern of communities is complicated by the presence of sub-habitats. Creeks, shallow depressions (salt-pans) and other surface features cause local varieties in environmental factors.

(*b*) *Adaptations to the habitat*.

(*i*) *Resistance to salinity*. Halophytes are adapted to cope with the stresses imposed by salinity (*see* VI).

(*ii*) *Reproduction*. Annuals are rare except in the narrow pioneer zone which is kept open by tidal action. The majority of salt marsh plants are shortlived perennials. This strategy seems to be the best compromise. The annual habit has the disadvantage of having to become re-established each year. The longlived

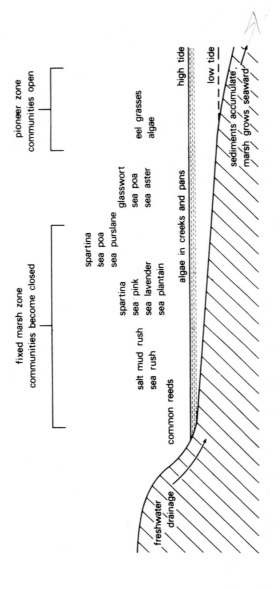

FIG. 52 *Generalised succession on a salt marsh.*

perennial habit does not give the opportunity to change location frequently in response to changing conditions as the marsh grows.

(*iii*) **Resistance to mechanical drainage.** Plants must be able to tolerate the force of tidal action. Many have narrow, smooth leaves which present little resistance to water movement. Some algae have gelatinous secretions which bind the plant into the mud.

30. Salt marsh heterotrophs. Bacteria and fungi are the most important consumers of energy in salt marshes. Large herbivores are lacking. The mud supports abundant invertebrate animals including snails, mussels and worms. These all have euryphagic habits and exploit a wide range of food all year. The invertebrates form a food supply for birdlife.

31. Ecosystem function.

(*a*) **Productivity.** Salt marshes are among the most productive ecosystems in the world. In some cases primary productivity is between 1000 and 2500 Kcal/m^2/year. These high rates are achieved because the marshes contain abundant nutrients. Salt marsh plants expend a lot of energy in respiration required to power the halophytic adjustments (*see* VI). In spite of this, net primary productivity is still in the order of 1.4 per cent of incident light energy. The standing crop of perennials is augmented by the growth of algae which may form 25 per cent of the total plant biomass.

(*b*) **Food chains.** Herbivores are scarce in the ecosystem. Most autotroph biomass is consumed by bacteria and fungi so that the flow of energy is short circuited. Food chains are reduced but may be complex because of the wide niches. The rate of energy flow through the system is high.

(*c*) **Nutrient cycling.** The salt marsh acts as a trap for nutrients brought by inland drainage and the tides. The abundant nutrients are cycled quickly. Decomposition in marshes is poorly understood but does seem to be aided by the mechanical breakdown by tides. Similarly, nutrients are moved throughout the marsh by tidal action. Dead grasses may be decomposed slowly but algae are broken down quickly to give a continual turnover of nutrients.

Some salt marsh plants experience deficiencies of trace elements. This is due to two reasons.

(*i*) Anaerobic conditions cause sulphates to be changed to

insoluble sulphides so that nutrients held in these compounds, such as iron, are unobtainable.

(*ii*) The presence of salt causes ion antagonism, that is, the uptake of some nutrients is inhibited by the presence of others. Many halophytes have low requirements for iron and manganese as an adaptation to these conditions.

SAND DUNES

Sand dunes form on coasts where large expanses of sand are exposed at low tide and where the shoreline topography is gentle. Wind blown sand is stabilised by vegetation and accumulates to form dunes.

32. Environmental conditions.

(*a*) *Sand mobility*. The dune habitat is characterised by shifting sand. This may bury plants, expose roots, cause abrasion and change the surface relief thus altering the microclimates.

(*b*) *Temperatures*. Sand is a poor conductor of heat. Surfaces are exposed to extremes of temperature but the sand acts as a buffer minimising variations in temperature within the dune.

(*c*) *Water supply*. Dunes are extremely arid because surface water drains away quickly.

(*d*) *Wind velocity*. Sand dunes are exposed to onshore winds. These accentuate the problems of desiccation and cause mechanical damage to plants.

33. Sand dune autotrophs.

(*a*) *Vegetation communities*. The flora contains a proportion of characteristic species but it is not as specific as that of salt marshes. Consequently it is a much richer flora. The vegetation includes a large number of species with different degrees of tolerance to sand covering and exposure to environmental extremes. These form various communities related to the distance from the sea and the age of the dune. The communities can be viewed as a succession sequence from bare sand on the beach to woodland inland.

(*b*) *The dune sequence*. The succession proceeds through three main phases (*see* Fig. 53). These may not all be present in one location because of erosion, habitat change or land-use.

(*i*) *Fore dunes or embryo dunes* are formed by low accumulations of sand near to high water level. Typically they are colon-

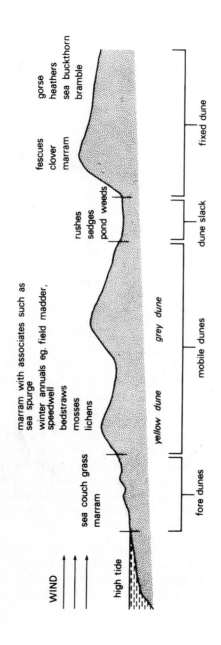

FIG. 53   *Generalised succession on sand dunes.*

ised by sea couchgrass (*Agropyron junceiforme*) and marram grass (*Ammophila arenaria*). Vegetation is sparse.

(*ii*) *Mobile dunes* still have large areas of bare sand. Marram grass is the main coloniser. This has extensive root systems, can stand desiccation and grows up through the sand as it is covered. For these reasons marram is the principal agent in stabilising the dunes.

Mobile dunes may be divided into *yellow dunes,* which have little organic matter in the soil, and the older *grey dunes* which have a shallow layer of organic matter near the surface.

(*iii*) *Fixed dunes* have a complete vegetation cover. Sand is stabilised by the closed community. Marram, which needs a constant supply of fresh sand to grow, becomes less frequent. The vegetation is dominated by grasses, such as the fescues, and low growing herbs. This dune grassland may develop into dune scrub or heath. Ultimately woodlands colonise the area.

(*iv*) *Dune slacks.* Wind erosion in the mobile dunes may cause "blowouts" which reach down to the water table. These dune slacks often have small ponds or marshes in them. Vegetation is usually halophytic but may contain some freshwater species.

34. Sand dune heterotrophs. Large herbivores are scarce apart from rabbits which graze mainly on the dune pasture communities. If grazers are absent grasses and sedges grow more luxuriantly and shade out broad-leafed herbs. Invertebrates are prolific and diverse, particularly the arthropods. Worms may be common in the dune slacks. Birds form an important component of the system; they feed on the invertebrates and carry the seeds of many of the scrub species in from other areas.

35. Ecosystem function. Productivity is low because of the sparse vegetation and harsh environment. Food chains are short but may involve complex food webs. Small amounts of nutrients are cycled, especially in the early stages of colonisation. Decomposition occurs rapidly. Nutrients may be locked in plants for long periods in the later seral stages when perennials predominate.

36. Maintenance. The sand dune system is inherently unstable. The surface sward can be eroded easily by trampling or excessive grazing. Once the vegetation has been removed the ecosystem reverts to the mobile stages of development. In areas where sand dunes suffer pressure from recreational activities attempts are made to prevent sand movement. These include placing matting on the surface and planting marram grass to stabilise the sand.

BOGS

37. Introduction. Bogs occur in areas of poor drainage. The soil is acid and permanently waterlogged. These conditions promote anaerobic decay which is very slow. This process leads to the accumulation of organic matter forming deposits of peat, which may be several metres thick.

Most British bogs have been drained or destroyed. Some remain in western Scotland, northern England, Wales and certain places in southern England such as the New Forest.

38. Types. Bogs can be divided into three main types.

(a) *Valley bogs*. These occur in valleys and local depressions where there is some impediment to local drainage. The valley bog may have a stream flowing through it. This transports nutrients into the bog and may decrease the acidity along its course.

(b) *Raised bogs*. These have convex surfaces built up by the accumulation of peat. They may develop on valley bogs so that the depression is filled by the growth of the bog. The surface or *hocmorr* is gently domed from the centre but is bounded by a steep slope or *rand* at the edge. Raised bogs frequently have a water course or *lagg* round the periphery.

(c) *Blanket bog*. This develops on slopes of gentle gradient in areas of high rainfall. It is extensive in many of the upland areas of Britain such as the western Pennines, Dartmoor and western Scotland.

39. Bog vegetation

(a) *Bog mosses*. Sphagnum mosses are the most important dominant plant type, and include about 20 species in Britain. Each one has a definite tolerance limit to acidity and waterlogging so they are useful environmental indicators. The growth of the bog mosses induces acidity in the bog since the spagnum cell walls produce carboxylic acid. Dead biomass from the sphagna is largely responsible for the accumulation of peat.

(b) *Valley bogs*. Vegetation communities usually display concentric zonation in relation to the distance from the edge of the bog and the central drainage stream. Wet heath grows on the edge of the bog and merges into communities dominated by the bog mosses with purple moor grass (*Molinea caerulea*), sedges and bog plants such as the bog asphodel (*Narthecium ossifragum*) in association.

The central area may have a fen-like community, that is, one which likes nutrient-rich, alkaline conditions. Willow and alder grow to form a *carr* along the stream course.

(*c*) *Raised bogs*. Characteristically the surface has a series of hummocks and hollows. Communities are dominated by cross-leaved heath (*Erica tetralix*), bog myrtle (*Myrica gale*), bog mosses and sedges. The different species segregate in relation to the microrelief because of its influence on drainage (*see* Fig. 54). The bog surface grows by a series of microsuccessions in the hummock complex. Hollows become infilled by the accumulation of peat and eventually form new hummocks. Hummocks stop growing as they dry up above the water table and eventually form hollows between the new hummocks. In this way the whole surface is raised. The water table is raised with the surface because the peat and the growing mosses retain a lot of water.

(*d*) *Blanket bog*. The species are mainly the same as those of raised bogs but sphagnum mosses may not be as prolific. Deer sedge (*Tricophorum caespitosum*), cotton grass (*Eriophorum latifolium*) and purple moor grass predominate. Cross-leaved heath and bog myrtle are present throughout except in the wettest areas. Hummocks do not usually develop in a blanket bog.

40. Bog plant adaptations.

(*a*) *The problem of waterlogging*. Bog plants grow with their roots in permanently waterlogged ground. This presents problems for root respiration and the uptake of nutrients. Anaerobic decomposition produces acidity which may cause toxic conditions around the roots and prevent nutrient uptake. Bog plants have large intercellular spaces to facilitate the movement of gases round the body. Many require less oxygen for their metabolism than do comparable mesophytes. Leakage of oxygen from the roots may give local aerobic conditions and may help to overcome the accumulation of toxins.

(*b*) *The problem of nutrient deficiency*. Bogs are deficient in many plant nutrients because of the acidity and the storage of nutrients in peat. Three main strategies have evolved in response to this.

(*i*) *Nutrient accumulation*. Many bog plants, including the

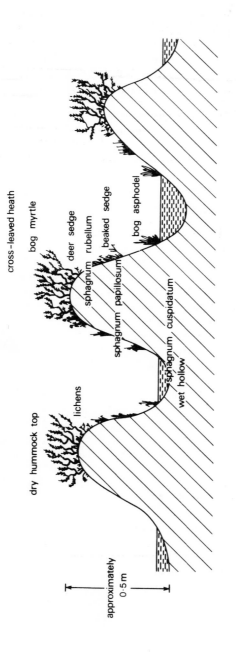

cross-leaved heath

bog myrtle

deer sedge

sphagnum rubellum

beaked sedge

bog asphodel

dry hummock top

lichens

sphagnum papillosum

sphagnum cuspidatum

wet hollow

approximately
0·5 m

FIG. 54　*Generalised variation in vegetation in relation to bog hummocks.*

sphagnum species, are adapted to grow under conditions of low nutrient supply. They can accumulate ions selectively from dilute solutions and retain them until they are needed for growth.

(*ii*) *Nitrogen fixation.* Nitrate is particularly scarce in bogs. Several plants, including bog myrtle, have root nodules containing symbiotic bacteria which enable them to fix atmospheric nitrogen.

(*iii*) *Carnivorous habit.* Some plants supplement their nutrient supplies by catching and digesting insects. British species of insectivorous or carnivorous plants include the sundews (*Drosera* species) and the butterworts (*Pinguicula* species). Both of these have sticky leaves which trap insects and then secrete enzymes to digest them.

41. Bog ecosystem function

(*a*) *Productivity and food chains.* Primary productivity is low compared with other ecosystems. Only the tundra and hot deserts are less productive than sphagnum bogs. Most net productivity is stored as the undecomposed organic matter of peat. Very little energy is transferred through food chains.

The detrital food chain is limited by the anaerobic conditions. In the grazing food chain invertebrates predominate, especially those species of insect which require water for reproduction. The invertebrates may form an important food supply for birds.

(*b*) *Nutrient cycling.* Decomposition is inhibited by acidity and the anaerobic conditions. Cycling occurs slowly because of the accumulation of nutrients in peat. Available nutrients are scarce although supplies may be augmented by inputs from precipitation and drainage from adjacent areas. If the input from drainage is significant, the bog is *rheotrophic*, whereas if the input is negligible the bog is *ombrotrophic*.

PROGRESS TEST 12

1. How did the climate of the Atlantic period differ from that of the boreal period? (**2**)

2. Why do deciduous woods in Britain often lack effective natural regeneration? (**5**)

3. What is the practice of coppicing? (**7**)

4. Why was the forestry commission formed? (**9**)

5. How may heathlands have originated? (**14**)

6. Why is ling the dominant plant on British heathlands? **(15)**

7. Which environmental conditions make chalk grasslands a distinctive habitat? **(20)**

8. Which features are typical of many chalk grassland plants? **(22)**

9. Why is the invertebrate fauna of chalk grasslands of special interest? **(24)**

10. How are salt marsh plants adapted to their habitat? **(29)**

11. Why do salt marshes have high rates of primary productivity? **(31)**

12. What are the principal environmental constraints in sand dune habitats? **(32)**

13. Outline the course of a typical succession on sand dunes. **(33)**

14. What are the differences between a raised bog and a blanket bog? **(38, 39)**

15. How do bog plants overcome the problems of nutrient deficiency? **(40)**

CHAPTER XIII

Practical Techniques

VEGETATION

1. Introduction. Description and analysis of vegetation usually involves taking a sample of the community because it is impossible to record all of the components. The techniques used and the size of the sample taken depend on the type of ecosystem being investigated and the information required. There are two basic types of vegetation survey.

(*a*) *Floristic methods* in which attention is given to the species structure of the vegetation.

(*b*) *Physiognomic methods* in which attention is given to the physical structure of the vegetation and to the life-forms of the plants present.

FLORISTIC METHODS

Information collected by these techniques can answer the following questions.

(*a*) Which species are present in the community?

(*b*) What is their relative abundance (i.e. in what proportions are they present)?

(*c*) What are the local distribution patterns of the species?

2. Quadrats. A quadrat is a wood or metal frame. It is placed on the ground and recordings are made for the area delimited by the quadrat. Both the size and the shape can be varied for particular tasks, for example circular and rectangular quadrats have been used. Conventionally the quadrat is 1 m^2. For diverse habitats like chalk grasslands a size of $\frac{1}{2}$ m^2 might be used whereas in homogeneous habitats like heathland, a 2 m^2 size might be used. Usually many quadrats are taken to obtain the information required.

3. Data collection in quadrats. Once the area to be sampled has been determined a number of observations can be made. These are recorded on a booking sheet (*see* Table XIV).

TABLE XIV. EXAMPLE OF A QUADRAT RECORD SHEET

Site: Walton Heath
Date: 21.6.80
Quadrat number: 3   *Size:* 1 m²

Species	Cover (Domin scale)	Density	Remarks
Calluna vulgaris	8	8	
Erica tetralix	3	4	
Galium saxatile	X	3	
Potentilla erecta	X	3	
Pteridium aquilinum	2	1	
Agrostis tenuis	1	4?	
Festuca ovina	1	5?	
Dicranium scoparium	X	1	
Cladonia arbuscula	X	2	
			Bare ground cover about 15 per cent

4. Species list. The names of all the species present are noted. These are usually grouped taxonomically, that is, plants in the same phylum, class and so on are listed together.

5. Cover. It is useful to know the relative importance of the species present. One of the ways to estimate this is to determine their relative cover abundance. Cover is usually taken as the area covered by the aerial parts of the plant but is sometimes taken as the basal area.

(*a*) *Subjective.* Earliest attempts to determine relative importance were based on subjective assessments and qualitative terminology such as rare, local and abundant. Scales were devised to make these assessments less subjective. The ones used most frequently are the Braun–Blanquet scale and the Domin scale (*see*

Table XV). However, these scales still rely on operator decisions and are therefore prone to error; also, they are non-linear.

TABLE XV. SCALES FOR ESTIMATING PERCENTAGE COVER

Cover	Domin scale	Braun–Blanquet scale
Cover about 100 per cent	10 ⎫	5
Cover 75–100 per cent	9 ⎭	
Cover 50–75 per cent	8 ⎫	4
Cover 33–50 per cent	7 ⎭	
Cover 25–33 per cent	6	3
Abundant cover about 20 per cent	5 ⎫	2
Abundant cover about 5 per cent	4 ⎭	
Scattered cover small	3 ⎫	
Very scattered cover small	2 ⎬	1
Scarce, cover small	1 ⎭	
Isolated, cover small	X	X

(b) *Quantitative.* The subjective element can be eliminated by accurate measurements of the cover of each species by using transparent graph paper or a network of fine crosswires. This approach is extremely time-consuming.

When the percentage covers of all the species in the quadrat are added the sum can exceed 100 per cent because the plants may overlap.

6. Density. This is a quantitative measure taken as the number of individuals per unit area, for example five dandelions per m^2. Its disadvantage is that it is not always possible to differentiate individual plants in the field. For instance, in the case of mosses and grasses it is hard to tell where one plant ends and the next begins.

7. Frequency. The chance of recording a certain species in any given quadrat is its frequency. It is noted as presence or absence in each quadrat for each species. Therefore, a species with a frequency of 50 per cent would occur in half of the quadrats taken.

8. Edge effects. When using quadrats the recorder faces the problem of what to do if a plant is at the edge of the frame. Part of the plant may be in the quadrat and part outside. This edge effect must be tackled consistently. The edge plants must either all be included or excluded from the record.

9. Point transects.

(a) *Method.* A line is taken through the vegetation and recordings are made along its length, usually at regular intervals. The point transect may be taken by a special bar (typically 1 m long) with sliding pins fixed at 10 cm intervals. The vegetation is sampled at the points where the pins touch the ground. Alternatively a measuring tape can be laid out and a pointer used to locate the sample positions. Point transects enable data to be collected quickly. They are particularly useful for studying relationships with environmental gradients.

(b) *Data collected.* Point transects provide a list of species present and a measure of their frequencies. The number of times each species is recorded indicates its frequency in the community. For example, a plant noted at 25 out of 50 points has a frequency of 50 per cent.

10. Sampling pattern. Quadrats and point transects may be taken either at random or systematically.

(a) *Random.* The samples are taken at random points located either by throw or by the application of random numbers (for example taken from Fisher and Yates Statistical Tables) to a grid superimposed on the area. Random sampling eliminates bias and produces a sample suitable for analysis by many statistical techniques. It may leave large parts of the habitat unsampled and may be unsatisfactory if relationships with environmental gradients are being investigated.

(b) *Systematic.* The sample is taken according to a definite pattern. For example, quadrats may be taken in parallel lines or laid end to end to form a belt. Point transects may be arranged downslope or in relation to other environmental factors. Systematic sampling ensures that undesirable gaps do not occur in the sampling scheme. The sample may not be suitable for analysis statistically.

In taking a systematic sample the recorder must take care to avoid inherent patterns in the habitat. For example, in a raised bog, hummocks may be spaced at 1 m intervals. If quadrats are taken at intervals of 1 m the same part of the hummock would be sampled each time thus giving a distorted impression of the community.

11. Plotless samples. This approach is used mainly in wooded areas where it is difficult to apply other techniques to measure density. Three methods are applied most frequently. All start from a point selected at random and are designed for use with large plants.

(a) *Closest individual method.* A measurement is taken from the random point to the nearest individual and this is then repeated.

(b) *Random pairs method.* This is an extension of the previous method in that an exclusion angle is set for 90° on either side of the sample point. The distance from the point to the nearest individual outside the exclusion angle is taken.

(c) *Point-centred quarter method.* Lines are erected from the random point at right angles to it giving four sectors in which the distance to the nearest individual is measured for each. The mean of these distances is calculated and taken as the figure.

12. Sample size. Generally the largest possible sample should be obtained. However, much will depend on the nature of the problem and the time available.

(a) *Minimum area.* The species area curve is a plot of the total number of species recorded against either the number of quadrats taken or the total area sampled. Quadrats may be taken at random or systematically (*see* Fig. 55(a)). Initially as a larger area is sampled there will be a sharp increase in the number of species encountered. With progressive increase in sample size fewer new species will be found so the curve will flatten (*see* Fig. 55(b)).

A point where the curve flattens can be taken as the minimum area necessary to sample to include the majority of species present in the community. The exact form of the curve will vary with the scales used therefore the point is usually taken where the curve approximates to some arbitrary relationship, for example a 5 per cent increase in species for a 10 per cent increase in area.

(b) *Normal distribution.* In samples of one species, sample size can be determined by measuring one aspect of the plant, for example petal size. When the data obtained approximate to a normal distribution curve this sample size is the one required.

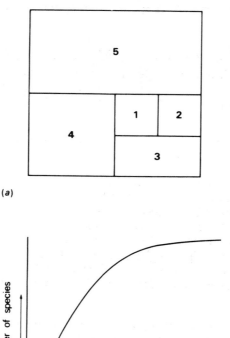

(a)

(b)

FIG. 55 *The minimum area required for a sample of vegetation. (a) Systematic arrangements of quadrats; (b) species–area curve.*

PHYSIOGNOMIC METHODS

13. Use of physiognomic techniques. Physiognomic techniques are often more useful than floristic ones in mapping and in understanding ecological relationships. They concentrate on the appearance and life-form of the vegetation. Information collected can answer the following questions:

(*a*) What type of plants are present in the community?
(*b*) What is the physical structure of the vegetation?

Vegetation structure does not necessarily change with species composition but does change with soils, topography and succession. Many methods of describing vegetation by its external appearance have been devised. Features used include colour, seasonality, layering, leaf-size and shape. Two schemes are described here as examples.

14. Raunkiaer scheme. In 1934 C. Raunkiaer devised a life-form classification based on the position of the perennating buds (*see* Table XVI). In his opinion the proportion of the different classes present in vegetation reflected the climatic conditions.

TABLE XVI. RAUNKIAER'S LIFE-FORM CLASSIFICATION

Life-form type	Description
Phanerophytes	trees and shrubs; branching system projects feely into the air
Chamaephytes	dwarf shrub, herbaceous perennials, cushion plants; perennating buds held in air at low level
Hemicryptophytes	have perennating buds in surface of soil
Geophytes	perennating buds beneath the soil surface on storage organ, e.g. bulb
Therophytes	annual plants
Hydrophytes	Equivalent to geophytes but have buds below the surface of the water

15. Dansereau's scheme. This scheme, devised by Dansereau in 1959, classifies vegetation on simple, easily observed distinctions between plants. In a sample line through the community each plant is first assigned to one of five primary categories based on life-form (*see* Fig. 56.) Subsequently it is assigned to categories for leaf shape and size, function and leaf texture. Symbols for each class can be used to represent the vegetation diagrammatically (*see* Fig. 57). This descriptive scheme conveys a visual impression of the vegetation and an indication of its function.

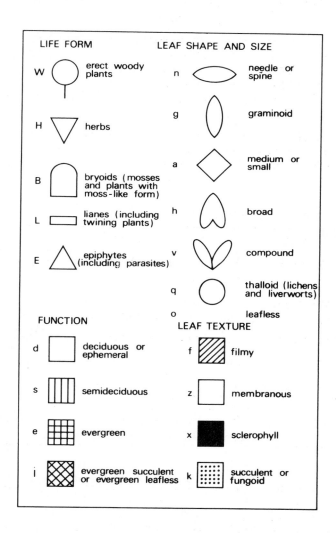

FIG. 56 *Key to the life-form symbols devised by Dansereau.*
(After P. Dansereau and J. Arros (1959), "Essais d'application de
la dimension structurale en phytosociologie" Vegetatio, **IX**, 48–49.

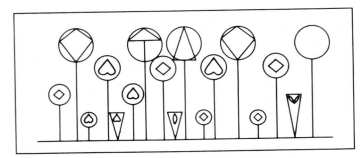

FIG. 57   *Example of a profile illustrating the use of the Dansereau life-form symbols.*

PRACTICAL TECHNIQUES FOR STUDYING ANIMALS

16. Introduction. Animals are more difficult to study than vegetation not only because they are mobile but also because many are nocturnal and vary their behaviour according to the time of year and the weather. Most animals are disturbed by the presence of an observer so their behaviour may be distorted.

It is difficult to obtain a representative and unbiased sample of animals. Some sections of a population may be more prone to capture than others, for example the old and infirm. Methods of capture may harm or kill the animals thus altering the structure of the population remaining.

17. Catching mammals and terrestrial invertebrates.

(*a*) *Small mammal traps.* The easiest method for capturing animals such as voles, mice and shrews is to use a standard mammal trap. This consists of two connecting small boxes having a total length of about 35 cm. One box, the entrance tunnel, is slightly smaller than the other, the chamber. The entrance tunnel has a trapdoor flap which can be lodged open. Animals are enticed into the trap by bait, and on entering the tunnel they touch a spring which closes the door. The traps must be set with adequate food and bedding to ensure the survival of the animal until it is collected.

(*b*) *Water traps.* These are useful for capturing insects which fly in daylight. The trap may be any shape providing that it is between 5 and 8 cm deep and about $\frac{1}{10}$ of a square metre in area. The inside is painted white or yellow and half filled with water containing a few drops of detergent. Insects are attracted by the colour of the trap and fall into the water.

(c) *Sweep nets.* Nets of fine mesh are mounted on rigid frames which are either triangular or semicircular in shape. The net is swept over long grass. At the end of each sweep the net should be turned so that the frame is vertical and the net hangs against it. This ensures that animals do not escape. The contents of the net should be transferred to collecting jars for examination.

When sampling an area a series of sweeps should be made as invertebrates tend to move downwards when disturbed. Successive sweeps should be lower until no more animals are caught.

(d) *Pitfall traps.* Any shallow jar sunk into the ground will serve as a pitfall trap. Crawling animals, such as woodlice and centipedes, will fall into the trap. The top of the trap should have a canopy of bark or straw to act as camouflage and protection against rain. Large containers should not be used as these will enable frogs and hedgehogs to reach the catch and eat it.

(e) *Suction traps.* Various designs of suction trap are available. They may be used for sampling the aerial populations of flying insects or for capturing small invertebrates from surfaces.

The most simple suction trap or *pooter* consists of a stoppered jar with two small tubes entering it. The end of one tube is placed over the insect to be caught; by sucking the end of the other tube gently the collector moves the animal into the jar.

Sophisticated designs include fans which suck air into the trap at a constant rate; the air escapes through a metal gauge cone. In these instruments the collecting jars usually contain alcohol so that specimens will be preserved while sampling takes place over a long period.

(f) *Light traps.* These are used for capturing night-flying insects. The most popular design is the Rothamsted light trap which incorporates a 200 watt bulb and a killing jar below the light source. Insects are attracted by the light and are killed by toxic fumes emitted by plaster of paris and tetrachlorethane in the killing jar.

18. Methods for extracting soil fauna. The soil ecosystem is so rich in animals that even casual sorting will reveal many life-forms. The numbers and sorts of animals present vary with the depth and type of soil. Comparisons can be made both between soils from different places and between the different horizons of one soil. Organisms may be abstracted by various techniques.

(a) *Mechanical* methods. These techniques are passive in that they do not require the animals to move. Consequently they extract inactive as well as active stages of the lifecycles. They extract dead animals as well as living ones and so may give an overestimate of population size.

(i) *Dry sieving* is used when the size of animals studied is significantly larger than the soil particles, for example beetles and earthworms. Different-size groupings of animals may be separated out by using a number of sieves of different mesh sizes.

(ii) *Wet sieving* is an efficient method for extracting medium and small-sized litter-dwelling organisms. Leaf litter is supported on a coarse mesh in a tank which is filled with water. When the litter is stirred the animals become detached and sink to the bottom of the tank where they can be collected.

(iii) *Flotation* enables a variety of animal groups to be extracted simultaneously. It is especially efficient in the separation of small arthropods which cannot be extracted by sieving. The method involves two operations. Firstly, the mineral soil is separated from the organic matter by washing and sieving. The different specific gravities of the two components cause them to settle from suspension at different rates. Secondly, the animals are separated from the plant material by differential wetting. The cuticles of arthropods are not wetted by water and so will collect on the surface of a tank whereas the plant material will sink to the bottom.

(b) *Behavioural methods.* These techniques rely on the movement of organisms in response to stimuli. For instance most species will move away from heat, light and desiccation. Inactive stages such as eggs, and dead organisms, will not be extracted.

Species vary in their tolerances to environmental factors and so will react differently. Therefore, the efficiency of extraction from one soil sample will vary with the type of animal.

The most frequently used extractor is the *Berlese–Tullgren* funnel. The sample is placed on a wire gauze in the mouth of a steep-sided funnel and is heated gently from above, usually by a light bulb. Organisms move down through the sample as it is heated and dried. Eventually they are forced out through the gauze into a collecting jar beneath.

19. Catching aquatic animals.

(a) *Water nets.* These are similar to the sweep nets except that the mesh is coarser and the frame more robust. Water nets may be used to collect any small aquatic animals.

(b) *Plankton samplers.* Both phytoplankton and zooplankton may be collected in a plankton net. This resembles a water net but has a small glass collecting vessel at its end. The net is dragged through the water slowly with the opening vertical. The sample may be transferred to other jars for storage.

20. Data collected. The data collected when sampling animals will depend on the problem under investigation. Species lists can be compiled for different habitats. Population sizes can be estimated (*see* 21(*a*) and (*b*)). Individual populations may be studied for their variability in size and colour, and to assess their age–sex structures. Lifecycles may be determined. By marking animals and sampling over several seasons information about movements can be collected.

21. Estimating population size. It is impractical to try to count the entire population of a species. A technique must be employed to estimate population size. The method used must be appropriate to the size, habitat and behaviour of the animals. Two of the most frequently used methods are described here.

(*a*) *Mark and recapture method.* In this the total size of a population is estimated by repeated sampling over time for marked individuals. A number of animals are collected and are marked (for example, with paint, nail varnish, tags or clipped fur). The size of the initial sample will vary with the species concerned and the size of the area.

The sample is released over the area uniformly and is left to mix with the rest of the population. A second sample is taken and the number of marked individuals recaptured is noted. The total size of the population (*N*) can be estimated by the relationship;

$$N = \frac{\text{number in sample 1} \times \text{number in sample 2}}{\text{number of marked individuals recaptured}}$$

This method makes a number of assumptions.

(*i*) That the marking method will not alter the behaviour of the animal, for example make them slower and more prone to recapture.

(*ii*) That the marked individuals will disperse through the area and mingle with the entire population at random.

(*iii*) That adequate time is left between the sampling for dispersion to take place.

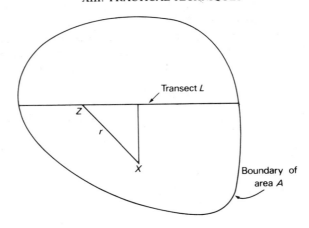

FIG. 58   *The linear transect method of estimating population density.*

(*iv*) That no immigration or emigration of individuals takes place across the boundaries of the area studied.

(*b*)   *Linear transect method.* This approach is useful for animals which are difficult to mark or capture. The population size is estimated from observations taken on a linear transect of length L (*see* Fig. 58). The transect is placed randomly through the area but should be away from the margins to avoid edge effects. The recorder walks along the transect. When an individual is spotted, for example a deer, the distance from the observer's position (Z) to the animal's position (X) is measured. This distance is r. There is a constant λ representing the visibility of the animals, which is characteristic for a species and type of habitat.

The total population (N) in an area of size A can be estimated from the relationship:

$$N = \frac{n(2n-1)A}{2L\Sigma r}$$

where n is the number of individuals seen.

The variance of the estimated density (that is the average amount of deviation from the mean) is:

$$\text{variance } N = \left[\frac{n}{(2L/A\lambda)} \right] \left[\frac{3n-2}{2(n-1)(2L/A\lambda)} - 1 \right]$$

where λ is estimated from $(2n - 1)/\Sigma r$.

The method assumes that individuals are dispersed randomly and that the probability of seeing individuals is equal on both sides of the transect.

PROGRESS TEST 13

1. What is the main difference between floristic and physiognomic methods of vegetation analysis? **(1)**

2. What are the disadvantages of the Domin scale of cover abundance? **(5)**

3. Describe the use of point transects. **(9)**

4. What is systematic sampling? **(10)**

5. What are plotless samples? **(11)**

6. What are the advantages of physiognomic methods of vegetation analysis? **(13)**

7. Describe three methods of catching terrestrial invertebrates. **(17)**

8. How is a Berlese–Tullgren funnel used? **(18)**

9. What assumptions are made in the mark and recapture method of estimating population size? **(21)**

CHAPTER XIV

Man as an Ecological Factor

INTRODUCTION

1. Man and ecosystems. Although humans form only a tiny percentage of the earth's biomass, man is the dominant species. At the beginning of the Pleistocene (*see* Table 9, VIII) men were subject to environmental controls within ecosystems in the same way as all animals are. Since then humans have progressed to become capable of altering their physical environment.

Some ecosystems have been modified or destroyed completely while new ones have been created. The distributions and numbers of species have been changed both intentionally and accidentally. Large areas have been subject to pollution. These activities have caused the extinction of many species and have disrupted ecosystem stability.

2. Population increase. The impact of humans on the biosphere results both from increasing technology and from the increase in numbers. World population growth has accelerated dramatically this century (*see* Fig. 59) because of the progressive reduction in mortality rates. The vast increase in human populations has accelerated the depletion of the world's resources and the destruction of its natural habitats.

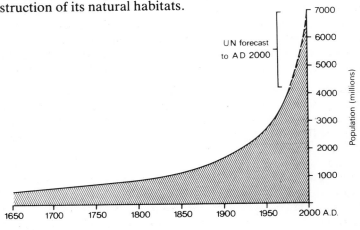

FIG. 59   *World population increase since 1650.*

262

DESTRUCTION OF HABITATS

The most obvious influence of humans on natural ecosystems is habitat destruction. This has resulted mainly from the pressures of agriculture, industrialisation and urbanisation but also from the exploitation of resources such as forests.

3. The spread of agriculture. The demand for agricultural land has destroyed many habitats. Bogs and fens have been drained causing the disappearance of their associated flora and fauna. Similarly, in other areas, irrigation has destroyed xerophytic communities. Many of the major forest formations have been cleared for agriculture, especially in temperate areas.

4. Deforestation. Exploitation of the world's forest reserves is increasing, reflecting the increasing demands for wood and paper pulp. Although some replanting takes place it is mainly in temperate areas and does not keep pace with felling. Removal of a forest cover causes disruption in nutrient cycling and hydrology. The changes in interception (*see* V) alter the amount of runoff. This in turn leads to problems of leaching and soil erosion. The increased sediment load in the water courses disrupts aquatic ecosystems.

The destruction of tropical rain-forests causes most concern because these complex ecosystems cannot regenerate once removed. Vast areas of tropical forests are being destroyed. Removal of the tree canopy not only causes the destruction of the habitats together with the organisms in them but also may have global repercussions. Forests are essential for the cycling of carbon and oxygen. Large-scale destruction may alter the balance of gases in the atmosphere. Similarly the change in the surface *albedo* (reflectivity) may disrupt the earth's radiation balance.

5. Land scarification. This general term is used to describe disturbances created by the extraction of mineral resources such as in quarries, sand pits and mine tips. Open-cast mining causes some of the most devastating alterations to ecosystems since the surface cover is removed completely. The waste heaps from deep mining are often toxic, and their surfaces are exposed to environmental extremes. It may take over 50 years for them to be colonised by vegetation. Less than 1 per cent of Britain is derelict due to scarification, but the amount is increasing and the distribution is uneven. For example, china clay working has produced over 6500 hectacres of derelict land in Cornwall. Sand

and gravel extraction is now the largest extractive industry in Britain. The results of this activity are widespread because of the nature of the deposits and the high cost of bulk transport. Although the workings destroy habitats the pits may fill with water and provide a refuge for aquatic species.

6. Urbanisation. The growth of the world's cities is increasing in relation to increasing populations and to increasing industrialisation. In Britain, urban areas now cover over 11 million hectares out of a total land area of 138 million hectares. The ecological impact of a city extends far beyond its fabric. Noise, pollution and the recreational activities of the inhabitants interfere with adjacent habitats. Ecosystems are destroyed as the city is built but these natural habitats are replaced with man-made ones which some organisms can exploit (*see* **10**).

ECOSYSTEM MODIFICATION

7. Overgrazing of natural pasture. The continued use of natural pasture without replenishment of nutrients is a feature of most pastoral economies. In many areas overgrazing has occurred as a result of economic and social incentives. The destruction of the natural vegetation leads to soil erosion and a reduction in the carrying capacity of the system. For example, the urge to make a quick profit led to overstocking of the western ranges of the United States between 1880 and 1930. This reduced its carrying capacity by about half.

Currently the problem of overgrazing is serious in underdeveloped intertropical areas. In many African pastoral communities a man's importance depends on the number of cattle he owns irrespective of their condition. Overstocking destroys the natural vegetation and accentuates the problems of drought in these areas.

8. Agricultural ecosystems. Farming may involve the modification of an existing ecosystem in, for example, the management of pastureland, or the replacement of a natural habitat with a "man-made" ecosystem. In both cases simplification takes place. In pastoral systems a single species of herbivore is maintained. In arable systems single types of crop are grown.

The aim of agriculture is to maximise the net productivity of desirable organisms. Unwanted species are regarded as weeds or pests and are excluded. A reduction in species diversity results, leading to a shortening and simplification of food webs. This type of system is inherently unstable and can be maintained only with a high input of energy (as for instance by weeding and harrowing) and materials (such as fertilisers).

9. Forestry. Selective felling and management of forests alters their species composition and physical structure. For example, in the cases of coppicing and pollarding (*see* XII, 7) management influences the functioning of the whole ecosystems.

CREATION OF NEW HABITATS

In destroying ecosystems man inevitably creates new habitats. These are usually less favourable for life but in some cases thay can form refuges for survival. Occasionally species adapt well to the new conditions and become more prolific.

10. Urban areas. Towns and cities contain diverse habitats for organisms. For instance, plants and animals may inhabit parks, gardens, sewers, waste tips and even houses. Common urban birds include the house sparrows (*Passer domesticus*), pigeons (*Columba livia*) and starlings (*Sturmus vulgaris*). The second species was originally a cliff dweller and has adapted to roosting on concrete ledges. Recently gulls have adopted the habit of scavenging on refuse tips. This can cause health hazards if decomposing material is carried by the birds to reservoirs.

Urban areas provide habitats for many mammals including foxes, hedgehogs, and mice. However, rats are the most important mammalian scavengers in towns. Most live in the sewage disposal systems possibly moving to the surface to feed at night. It has been estimated that 1 km of sewer can support over 500 rats. Rats act as the carrier of many diseases including the bacterium which causes *leptospirosis*.

Pet dogs and cats in towns form a significant reservoir of mammalian parasites such as fleas, tapeworms and roundworms. Many of these are transmitted easily between individuals and may also infect humans.

11. Wasteland. Sites such as disused industrial land, marginal farmland and disused airfields present open habitats for pioneer species. These types of area undergo secondary succession, providing refuges for many heliophytes excluded from agricultural land by farming techniques.

12. Hedgerows. These provide an important refuge for wild life requiring a woodland type of habitat, and form vital links for migration routes between woods.

The hedgerows may be ancient remnants of woodland (assart hedges) especially if they occur on parish boundaries. Even if the hedgerow has been planted relatively recently it may have a varied biota, particularly in herbaceous plants, birds and insects. For example, over 500 species of plant occur in hedgerows although only half of these are regarded as typical hedgerow plants. Very few of our rarest species are confined to hedgerows, but the alternative habitat space for the woodland plants is decreasing.

Many hedgerows are being removed because of the cost of their maintenance, the trend towards bigger fields and the idea that hedges reduce crop productivity due to shading and root competition. However, it is necessary to preserve some of them not only for their habitat value but also because they help to prevent soil erosion and ameliorate local climates.

13. Canals. The main canalways of Britain were constructed between 1750 and 1830. The network of waterways formed a continuous system of freshwater habitats which were cleaned regularly and so kept open for colonisation. This had a profound effect on the aquatic flora and fauna.

The potential for migration was enhanced by the movement of barges. Fragments of plants were transported in both directions along canals whereas in rivers movement was limited to the downstream direction unless the plant was moved upstream by an animal. The canal banks and towpaths provided open habitats for heliophytes.

The use and maintenance of canals has declined over the last century with the loss of many aquatic habitats.

14. Reservoirs. The increasing demand for water supplies has necessitated the construction of many reservoirs. Although these submerge terrestrial habitats new aquatic ones are created. The reservoirs are important for birdlife, particularly as resting places for migrants. The fluctuating water levels in most reservoirs make them ideal habitats for mud ephemerals such as red goosefoot (*Chenopodium rubrum*) and common mudwort (*Limnosella aquatica*).

15. Railways. The railway network of Britain reached its maximum extent about 1928 when over 8450 km of track were in operation. The railways themselves provided open habitats on the tracks and banks which were often burnt and so kept at early seral stages. The fabric of bridges, platforms and halts presented cool, moist habitats for ferns and mosses.

The routes formed continuous habitats for the spread and migration of plants and animals. Railway land enjoys some protection from public access and therefore gives a refuge for many birds, such as the nightingale.

Railway construction altered adjacent habitats. Some wetlands were drained and others created by the presence of embankments. New slopes on the sides of cuttings altered microclimates. South-facing cuttings provided warm habitats because of the increased insolation.

16. Roadways. Early unpaved tracks provided extensive areas of trampled and disturbed ground for the colonisation and spread of heliophytes. In the nineteenth century road surfacing techniques improved so that the majority of roads were paved.

Road verges became a sanctuary for wildlife. Until recently they were free from human disturbance except for grazing and occasional cutting. It is estimated that 420,000 hectares of England and Wales are road verges. This is over three times the area of the nature reserves. More than 700 species of flowering plant inhabit road verges. In addition this habitat is important for insects, small mammals and birds which rely on these animals for food.

In recent years the road verges have been subject to more disturbance from cutting, herbicides and construction for cables and pipes. Motorway verges are becoming important habitats as they are relatively undisturbed by humans. Populations of kestrels have increased in the vicinity of motorways reflecting the increase in small mammal populations on the verges.

MAN AS AN AGENT OF DISPERSAL

17. Altering the natural distributions of plants and animals. Man has altered the natural distributions of plants and animals both intentionally and accidentally. In both cases species have increased their population sizes greatly as a result of this intervention. It is estimated that two-thirds of the world's weeds and pests have become so as a result of their having been introduced into new regions by man. An organism introduced to a new environment is frequently presented with favourable conditions for expansion.

(a) There may be a total lack of natural enemies or diseases.

(b) There may be a vacant ecological niche in the new ecosystem which the species can exploit.

(c) Simplified agricultural "man-made" ecosystems may provide ideal continuous habitats.

The speed and extent of man's influence on distribution patterns have increased in relation to the spread of agriculture, urbanisation and travel.

DOMESTICATION

18. The neolithic revolution. Man began to domesticate wild plants and animals about 11,000 years ago. In the old world this cultural revolution started in the "fertile crescent region" of the Middle East extending through Palestine and Syria. Ancestors of our wheat and barley were used as arable crops. Similarly, other wild plants of the region such as flax, plum and carrot were used as food.

As the neolithic revolution spread to new areas some plants were adopted for use from the fertile crescent area. Other local native plants were also selected and bred for cultivation. For example, in south-west Asia crops included millet, radish, peach and lemon, whereas in India and south-east Asia rice and banana were used. Domestication of some animals may have occurred before that of plants. Evidence suggests that earlier cultures had domesticated the wolf and jackal for use in hunting. It is likely that many animals, such as sheep and goats, were domesticated by about 6000 BC soon after the first cultivation of plants.

The domestication of plants and animals has had a profound effect on the distribution of species. Desirable types have been bred and taken to new areas as agricultural knowledge spread.

19. Organisms following the spread of agriculture. The spread of crops was accompanied by the spread of weeds and pests. The simplified agricultural ecosystems allowed the invasion of plants requiring open habitats and provided favourable conditions for many insects.

Some weeds extended their ranges in parallel with a particular crop. For example gold of pleasure (*Camelina sativa*), a native of eastern Europe, is a weed of flax fields. The seeds of the two species are of similar size and are difficult to separate by winnowing. The ranges of the weed and the crop were extended simultaneously by human action.

Other weeds were already present in regions as incidental species often being relicts of heliophytic communities. The arrival of agriculture presented the opportunity for these species to multiply and extend their habitat space. British examples include the mugwort (*Artemisia vulgaris*) and the mayweeds (*Matricaria* spp.).

INTENTIONAL INTRODUCTIONS AND ESCAPES

20. Plants. Man has taken trees, shrubs and flowers from one region to another for their ornamental value. In Britain many have become naturalised and have established themselves as part of the flora. For example sycamore (*Acer pseudoplatanus*) was introduced by the Romans. More recently the hottentot fig (*Carpobrotus edulis*) was imported from South Africa. It escaped from seaside parks and gardens and now grows in abundance on sandy cliffs along the south coast.

21. Animals. Many animals have been introduced directly into the wild in new regions or have become established following escapes from captivity.

(*a*) *Direct introductions.* A well-documented case is that of the American grey squirrel (*Sciurus carolinensis*) which was introduced to a number of sites in the British Isles in the nineteenth century (*see* Fig. 60). The species spread rapidly between 1920 and 1925 following a population decline of the native red squirrel (*S. vulgaris*). The grey squirrel is the stronger competitor in deciduous woodlands and has become the more abundant species.

(*b*) *Escapes.* In Britain several species have escaped from fur farms and have become naturalised. The large South American rodent, the coypu (*Myocastor coypus*), has become established in East Anglia. It causes little harm apart from the destruction of some reed beds. The mink (*Mustela lutreola*) has been recorded in every British county. It inhabits freshwater systems and eats a

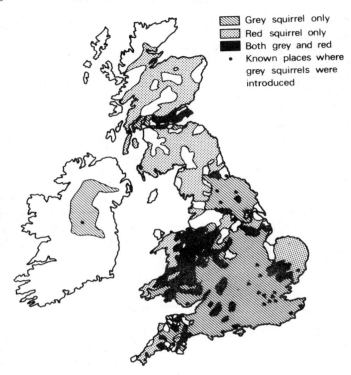

FIG. 60   *Map of squirrel distribution in Britain. (After Britain, Sunday Times Publications).*

variety of prey. Unfortunately its niche brings it into direct competition with the native otter.

Several species of deer have escaped from parklands, including the Chinese water deer and the muntjac. Both of these species are small, being less than 60 cm tall, inhabit dense undergrowth and have become established in many counties.

22. Accidental introductions. There are many well-documented cases of the accidental introduction of plants and animals to new areas. Most of the organisms involved have become pests in their extended distribution range.

(*a*) *The brown rat (Rattus norvegicus)* has been transported to many countries on ships. It has caused havoc in the unstable ecosystems of oceanic islands where it has depleted the less competitive endemic fauna.

(b) *Many weeds* have spread across continents as a result of the movement of people and foods. For example, the seeds of the Thanet cress (*Cardaria draba*) were transported across Europe accidentally in the hay stuffing of mattresses in the early nineteenth century. This weed species is now present in most English counties and is difficult to eradicate.

(c) *Diseases.* Man has inadvertently introduced diseases to new areas, as in the case of the introduction of chestnut fungus (*Endotheria parasitica*) from Asia to the eastern United States early this century. The disease is not a killer in its own environment in Asia, but in America the chestnut trees have almost been wiped out by this alien parasite. Similarly the fungus-beetle combination which causes dutch elm disease was introduced to Britain in imported elm timber. The disease has decimated the elm trees of this country.

AIR POLLUTION

23. Types and trends. Substances are pollutants if they cause detrimental effects. Many chemicals which can be pollutants, for example sulphur, are needed by plants and animals in small amounts. Airborne pollutants include particulate matter (both solid and liquid particles) and gases such as sulphur dioxide, hydrogen fluoride and nitrogen oxides. Most air pollution derives from urban areas, industrial processes, power generation and transport. In Britain emissions of smoke (particulates) have decreased markedly since 1956 when the Clean Air Act was passed (*see* Fig. 61). However sulphur dioxide (SO_2) emissions, which were not controlled by the Act, have increased steadily. This pollutant is produced when fossil fuels are burnt and it remains the most important air pollutant in the country.

24. Dispersal. Air pollutants are dispersed by the wind. Large particles are deposited near the source by gravity. Smaller particles and gases may undergo *long-range transport* before being removed from the atmosphere. During this time the *primary pollutants* may undergo chemical reactions to produce *secondary pollutants.* In areas of intense sunlight photochemical reactions take place. Nitrogen oxides and hydrocarbons react to

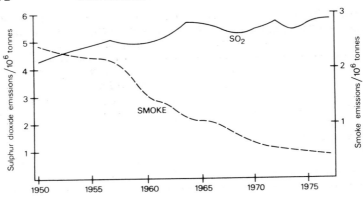

FIG. 61   *Recent trends in smoke and sulphur dioxide emissions in Britain.*

form ozone (O_3). This may combine with other chemicals to form complex pollutants.

Air pollution can travel between countries, or even between continents, before being removed from the atmosphere. Ultimately it is either absorbed by clouds (*rainout*), by precipitation (*washout*), by vegetation, by soil, by water bodies or by artificial surfaces such as buildings.

25. The effects of air pollution on plants.

(*a*) *Metabolism and structure.* Particulate matter may block stomata inhibiting gaseous exchange. Sulphur dioxide interrupts the plants' physiology in many ways. For example, when converted to sulphite (SO_3) it interferes with photosynthesis. When converted to sulphuric acid (H_2SO_4) it causes changes in the pH balance of cells, so altering the permeability of cell membranes. Increased acidity leads to the breakdown of chlorophyll molecules. Similarly fluorides and photochemical pollutants cause disintegration of leaf tissue. These effects may kill the plant or may reduce its productivity without any visible damage.

(*b*) *Changes in distributions.* Many plants are very sensitive to air pollution. The distribution of these species has altered as air pollution has spread. In Britain many lichens have become extinct near urban areas during the last two centuries. Around cities the number of species of lichen and bryophyte present increases with distance away from the city centre (*see* Table XVII).

TABLE XVII. THE DISTRIBUTION OF LICHENS AND BRYOPHYTES IN RELATION TO SULPHUR DIOXIDE CONCENTRATIONS WESTWARDS FROM NEWCASTLE-UPON-TYNE

	City centre	Inner suburb (2.5 km)	Outer suburb (8 km)	Edge of city (12 km)
Annual average SO$_2$ concentration (μg m^{-3})	200	160	90	65
Number of species	18	20	40	58

Many fungal parasites are killed by air pollution. For instance, the black spot fungus of roses (*Diplocarpon rosae*) cannot tolerate high concentrations of sulphur dioxide.

26. The effects of air pollution on animals.

(*a*) *Direct effects.* The combined action of particulates and acidic gases has resulted in many dramatic incidents of air pollution damage to urban populations (*see* Table XVIII). Inhalation of severely polluted air produces irritation of the respiratory tract, which induces inflammation of the lung, coughing and heart strain. The elderly and infirm may die because of this.

People in contact with polluted atmospheres at work such as miners and quarry men, are liable to develop diseases caused by "industrial pollution". These include silicosis, pneumoconiosis, anthracosis and emphysema. Severe irritation of respiratory tissue by particulates may cause irreversible disintegration of the lung.

Photochemical smog causes irritation of the eyes and respiratory track. Some of the compounds involved are thought to be carcinogenic (cancer-producing) but no fatalities have been ascribed to this type of pollutant yet.

Other pollutants emanating from individual processes have localised but important effects. For instance, fluorides, produced by brickworks, aluminium works and steelworks, are responsible for *fluorosis*. Livestock which have eaten grass contaminated with fluorides develop brittle bones and mottled teeth because fluoride replaces the calcium in the skeleton.

TABLE XVIII. EXAMPLES OF MAJOR AIR POLLUTION EPISODES IN URBAN AREAS

Date	Place	Excess deaths*
February 1880	London, UK	1000
December 1930	Meuse Valley, Belgium	63
December 1952	London, UK	4000
November 1953	New York, US	250
January 1956	London, UK	1000
February 1963	New York, US	200–400

* Excess deaths are those in addition to the normal death rate for the season.

(b) *Indirect effects.* Air pollution may operate as an agent of natural selection as in the case of the peppered moth (see VIII, 12).

Some changes in distributions could be due to air pollution but it is difficult to isolate the influence of this factor from the effect of others such as urbanisation and agriculture. As would be expected, invertebrates living on tree trunks exhibit the same general relationship with distance from city centres as do lichens. In particular, species grazing on lichens, such as the bark louse (*Reuterella helvimacula*), decline as the epiphytes decline and air pollution increases.

WATER POLLUTION

27. Types and trends. Water pollution has been studied more intensively than air pollution mainly because water forms a defined environment and is an economic commodity. Water pollutants include industrial wastes, fertilisers, sewage and natural seepage from mineral deposits.

Many of Europe's major rivers became devoid of life in the nineteenth century because of severe pollution. Attempts to rectify this situation have been successful in many places. For example, in London the Thames now supports over 97 species of fish whereas 100 years ago very few could survive. In contrast, marine pollution is increasing steadily particularly because of oil spillages.

28. Oxygen depletion. The oxygen content of water is low naturally. If water is polluted with organic waste, such as sewage, the oxygen content of water is depleted drastically. This occurs because organic material requires oxygen for decomposition. The daily sewage from one person needs approximately 9000 litres of water for oxidation. Some industrial organic wastes take even more. When water becomes deoxygenated the aquatic fauna suffocates.

The *biological oxygen demand* (BOD) of water is the amount of oxygen in milligrams per litre (mg/l) consumed within the sample over 5 days at 20°C in a dark closed container. This measure indicates the potential deoxygenation of the body of water.

29. Eutrophication. Water may become excessively enriched with nutrients. These can be either organic, such as in sewage, or inorganic as from inorganic fertilisers. The increase in nutrient supply favours the growth of algae, leading to "algal blooms" which shade out other plants. This change in the pattern of primary productivity alters food chains. Decomposing algae masses discolour and deoxygenate the water. Most fish are killed off and the entire ecosystem is disrupted. Eutrophication is most severe in lakes where flushing by fresh water takes a long time.

30. Detergents. Simple soaps do not cause water pollution problems because they are broken down easily by agents of decay. Synthetic "hard" detergents, manufactured after 1918, had complex molecules which resisted decomposition in sewage works. On entering water courses the detergents formed persistent masses of foam and lowered the oxygen absorption capacity of the water. This not only caused depletion of the fauna but also a health hazard as bacteria from the sewage were blown around on the foam. More recently "soft" detergents have been introduced. These have relatively simple molecules which can be broken down in sewage works.

31. Thermal pollution. When industrial cooling water or effluent enters a water course it is usually at a higher temperature than the river or estuary. The resultant warming causes thermal pollution. The rate of the metabolism of poikilotherms and the rate of decomposition increase in relation to temperature, so at higher temperatures the oxygen consumption increases, leading to deoxygenation. The species structure of the ecosystem may alter reflecting the different temperature regime.

32. Particulate matter. Particulates may enter water courses as a result of mining or agricultural activities. In suspension the particles increase the turbidity of the water. This reduces the amount of light penetrating for use in photosynthesis. Suspended particles are abrasive to fish gills and can interfere with filter feeders.

When deposited, the particles change the bottom environment of the water course. Silting in streams and rivers may destroy the spawning grounds of species needing gravel in which to lay eggs. In estuaries silt may cover rocky habitats needed for organisms such as mussels.

33. Dissolved salts. The concentration of dissolved salts in freshwater habitats can increase either because of irrigation on adjacent land or by inundation of seawater. Although the salts may not be directly toxic they upset the osmotic balances of organisms. Freshwater forms are adapted to live at low osmotic pressures. If the salt content of the water increases markedly water is drawn from the bodies of organisms by osmosis. This leads to rapid death in many species of fish.

34. Toxic substances. Poisons in water may not always derive from an artificial source. Hydrogen sulphide produced from decaying matter can be toxic. Similarly the excreta from dinoflagellates (a type of plankton) can be lethal.

Toxins introduced by human activities include the heavy metals, for example lead, zinc and mercury, phenols and cyanides. Most metallic poisons are not broken down in water, but accumulate in sediments where they can contaminate shellfish and enter the food chains.

PESTICIDES AND RADIOACTIVE FALLOUT

35. Pesticides and radioactive substances. The most toxic pollutants of the biosphere are pesticides and the radioactive byproducts of nuclear explosions. Both of these groups of substances have become extremely important since the second world war. Pesticides include the organochlorines, such as dieldrin and DDT, and the organophosphates, such as parathion and malathion. Biologically the most important radioactive substances are the isotopes strontium 90, caesium 137 and iodine 131, which fall to earth as dust after nuclear explosions.

Both pesticides and radioactive fallout include very stable chemicals which are slow to break down to harmless substances, and persist for many years once discharged to the environment. The substances involved may be present in minute concentrations in the air, soil or water, but can become concentrated through the food chains. For instance DDT residues can become progressively concentrated in tissues as they are passed from one trophic level to the next.

The degree of concentration depends on the type of animal tissue in which the pollutants are stored and the length of the food chain. Generally the top predators, such as eagles, accumulate the greatest concentrations. For this reason they are most susceptible to harm. In the case of birds, entire species may be eliminated by DDT because contaminated adults fail to produce viable eggs.

CONSERVATION

36. The need for conservation. It has been noted earlier in this chapter that human activities lead to simplification of ecosystems, disruption of food chains, destruction of habitats, pollution and interference with natural distribution patterns. All of these involve instability and the depletion of natural resources.

Biological diversity has been reduced both directly by extermination, through for example hunting, and indirectly, through for example the destruction of habitats and excessive competition between organisms in refuge areas. Once a population is reduced it becomes unstable. Simple ecosystems are less stable than complex ones and therefore the species in them are more prone to extinction thus impoverishing the ecosystem further.

37. The idea of conservation. Nature conservation is a very broad term. It aims to apply the principles of ecology to manage the entire environment taking into account all human activities to attain a better balance for varieties of life and scenery. Conservation is not static preservation but is active management based on a thorough knowledge of ecology.

38. The ecological history of conservation. The philosophical ideas giving rise to the practices of conservation are of ancient origin. Areas of land were reserved as natural habitats in India as early as 300 BC. However, the need to conserve was rarely expressed until the second half of the nineteenth century when it was realised that environments were deteriorating rapidly on a global scale. The industrial revolution caused destruction, and in developing countries animal populations were decimated, as for instance in the case of the Great Plains buffalo in the United States. The first national park was established in 1872 at Yellowstone partly as a refuge for the remnants of the buffalo herds.

Conservation gained further impetus in the 1930s when drought and agricultural depression in Texas and Oklahoma revealed overexploitation of soils, and vast dustbowls were created by soil erosion. As a remedy the US soil conservation service was established in 1934. This represents a major step in the history of conservation because it acknowledged the need for active management of a natural resource.

39. Pressures on world ecosystems. Many of the factors threatening world ecosystems are dealt with in detail earlier in this chapter but it is useful to summarise them in the context of conservation.

(*a*) *Increase in world population.* The vast increase in the population of the world forms the greatest threat to ecosystems. People require food and resources and so less habitat space can be left for other species.

(*b*) *Agricultural technology.* Increasing mechanisation and use of chemicals in farming results in pollution and simplification of ecosystems.

(*c*) *Industrialisation and urbanisation.* The spread of industry, and cities together with the general rise in living standards results in increased exploitation of resources for raw materials and power, and produces pollution.

(*d*) *Introduction of exotic species.* Accidental and intentional introductions of new species into communities change distribution patterns and disrupt stability.

40. Pressures on individual species. Most of the factors threatening world ecosystems apply also to single species but in addition several other factors may threaten single types of organism.

(*a*) *Habitat alteration.* Slight changes to ecosystems may eliminate some species because their niche is destroyed although the ecosystem as a whole may survive.

(b) *Collecting for zoos and research.* Until the late 1960s, many rare species were collected for zoos. This often included slaughtering many more animals than were required. For example, in order to collect orang-utans, which inhabit the forests of Borneo and Sumatra, adult females were killed so their young could be captured.

(c) *Hunting.* Many species have been brought to the verge of extinction by excessive hunting. In some cases killing was for sport or for a small part of the animal, for instance the skins of tigers and leopards and the feathers of the ostrich. In other cases hunting formed the basis for large industries, for example whaling. At the end of the second world war over 140,000 blue whales remained. This animal is the largest mammal ever to have existed. Now the population is less than 2000 because of the impact of over-exploitation especially by "factory-ship" methods.

(d) *Control of predators.* Carnivores have been selectively killed by men in the past to protect livestock and game. This practice has led to problems in some areas as predators act as a control on herbivore populations. When released from this control natural herbivores increase and compete with livestock for herbage.

41. Conservation policy. The idea of conservation encompasses two views which appear to be contradictory.

(a) It aims to preserve or protect wildlife and habitats.
(b) It aims to use resources and produce a sustained yield.

Conservation policies have the problem of reconciling these objectives. Humans need industry and agriculture to maintain their way of life, and so must use the world's resources. Natural evolution involves the constant extermination of species. Therefore the preservation of individual types of organism must be put into perspective.

Modern conservation aims to preserve in balance the maximum rates of biological productivity, energy transfer and nutrient cycling in all ecosystems while maintaining the quality of the environment.

42. Conservation methods. Effective nature conservation requires an attitude of mind which encourages man to adopt a constructive relationship with the environment. Four aspects are of special importance.

(a) *Protected areas.* Certain areas are protected from exploitation.

(i) *Nature reserves.* In these the balance of species is managed by conservationists. Some nature reserves in Britain cover only a few hectares. In contrast some, such as the African game parks, are extremely large.

(ii) *National parks.* Britain has ten national parks including Snowdonia and the North Yorkshire moors. These parks cover 9 per cent of the country and were established by the National Parks and Access to the Countryside Act of 1949. This legislation provided powers for "preserving and enhancing the natural beauty" of the designated areas.

(iii) *Other special areas.* In Britain certain areas enjoy other categories of protection: 27 *areas of outstanding natural beauty* have been identified, including the Cotswolds and the Sussex Downs; certain places have been identified as sites of *special scientific interest*. Both of these types of area are smaller than the national parks. It was assumed that they did not need positive management but that land-use in them should be strictly controlled.

(b) *Multiple land-use areas.* In some cases effective management may allow several activities to take place in one habitat. For example, forestry areas may not need to be reserved exclusively for timber production. They may be used for wildlife refuges, recreation and watershed management at the same time. Similarly moorland may be used for grazing, game, recreation and military exercises.

(c) *Protected species.* Certain endangered species need special protection by legislation. Examples in Britain include the Protection of Birds Act 1954, the Conservation of Seals Act 1968 and the Protection of Rare Plants Act 1974.

(d) *Pollution control.* In Europe and North America the most severe instances of air and water pollution have been controlled by legislation. However, much needs to be done on a global basis. In particular, the application of biological principles in pest control could reduce much pollution of soil and water. Pests could be eliminated by their natural predators or parasites in methods of *biological control* rather than by using chemicals.

43. Administration. Most developed countries now have authorities to administrate conservation efforts. In Britain the *National Trust* was founded in 1895 to preserve places of historic interest and natural beauty. *The Nature Conservancy Council* (the official research council concerned with wildlife in Britain) was established by Royal Charter in 1949. Currently the direction of national government agencies are being coordinated by international organisations like the *Conservation Foundation* and the *World Wildlife Fund*.

PROGRESS TEST 14

1. What have been the main causes of habitat destruction? **(3–6)**
2. How are agricultural systems different from natural ones? **(8)**
3. Why are urban areas important for wildlife? **(10)**
4. Why are transport systems important as habitats? **(13,15,16)**
5. How has the neolithic revolution altered the distribution of non-agricultural plants? **(19)**
6. Give an example of an alien animal introduced intentionally to Britain. **(21)**
7. Give an example of a species spread accidentally by humans. **(22)**
8. How may air pollution effect the distributions of plants? **(25)**
9. What is fluorosis? **(26)**
10. What causes oxygen depletion in water? **(28)**
11. What are the effects of thermal pollution in water? **(31)**
12. Why do pesticides have severe detrimental effects on ecosystems? **(35)**
13. What are the basic aims of conservation? **(37, 41)**
14. What are the main pressures on individual species? **(40)**
15. What is the multiple land-use policy? **(42)**

Selected Reading

GENERAL TEXTBOOKS WHICH COVER SEVERAL ASPECTS OF ECOLOGY

Anderson, J.M. *Ecology for Environmental Sciences: Biosphere, Ecosystems and Man.* (Edward Arnold, 1981)

Etherington, J.R. *Environment and Plant Ecology.* 2nd Ed. (Wiley, 1982)

Jones, R.L. *Biogeography: Structure, Process, Pattern and Change within the Biosphere.* (Hutton, 1980)

Pears, N. *Basic Biogeography.* (Longman, 1977)

Ricklefs, R. *Ecology.* 2nd Ed. (Nelson 1980)

Simmons, I. *Biogeographical Processes.* (George Allen and Unwin, 1982)

Smith, R.L. *Ecology and Field Biology.* 3rd Ed. (Harper and Row, 1980)

Tivy, J. *Biogeography: A Study of Plants in the Ecosphere.* 2nd Ed. (Oliver and Boyd, 1977)

PRODUCTIVITY

Hall, D.O. and Rao, K.K. *Photosynthesis.* 2nd Ed. (Edward Arnold, 1977)

Jones, G. *Vegetation Productivity: Topics in Applied Geography.* (Longman, 1979)

ENVIRONMENTAL FACTORS

Brady, J. *Biological Clocks.* (Edward Arnold, 1979)

Bryant, R.H. *Physical Geography Made Simple*, 2nd Ed. (W.H. Allen, 1979)

Hardy, R.N. "Temperature and Animal Life", *Studies in Biology.* No. 35. 2nd Ed. (Edward Arnold, 1979)

Lofts, B. "Animal Photoperiodism", *Studies in Biology.* No. 25. (Edward Arnold, 1970 (reprint 1978))

Strait, B.R. and Billings, W.P. "Vegetation and Environment", *Handbook of Vegetation Science.* (Dr W. Junk, The Hague, 1974)

Sutcliffe, J.F. *Plants and Water.* 2nd Ed. (Edward Arnold, 1979)

Sutcliffe, J.F. *Plants and Temperature.* (Edward Arnold, 1977)

SOILS

Bridges, E.M. *World Soils.* 2nd Ed. (Cambridge University Press, 1978)

Courtney, F.M. and Trudgill S.T. *The Soil: An Introduction to Soil Study in Britain.* (Edward Arnold, 1976)

Curtis, L., Courtney, F. and Trudgill S.T. *Soils in the British Isles.* (Longman, 1976)

Trudgill, S.T. *Soil and Vegetation Systems: Contemporary Problems in Geography.* (Clarendon Press, 1977)

POPULATION DYNAMICS

Krebs, C.J. *Ecology: The Experimental Analysis of Distribution and Abundance.* 2nd Ed. (Harper and Row, 1978)

Krebs, J.R. and Davies, N.B. *Behavioural Ecology: An Evolutionary Approach.* (Blackwell Scientific Publications, 1978)

Solomon, M.E. *Population Dynamics.* 2nd Ed. Studies in Biology No. 18. (Edward Arnold, 1976)

EVOLUTION

Dowdeswell, W.H. *The Mechanism of Evolution.* 4th Ed. (Heinemann Educational, 1973)

Ford, E.B. *Evolution Studied by Observation and Experiment.* (Oxford University Press, 1973)

Savage, J.M. *Evolution.* 3rd Ed. (Holt, Rixehart and Winston, 1977)

Scientific American. *Evolution.* (W.H. Freeman and Co., 1978)

SPECIAL NICHES

Scott, G.D. *Plant Symbiosis.* Studies in Biology No. 16. (Edward Arnold, 1969)

Wilson, R.A. *An Introduction to Parasitology.* 2nd Ed. Studies in Biology No. 4. (Edward Arnold, 1979)

MIGRATION AND DISTRIBUTION PATTERNS

Baker, R. *The Evolutionary Ecology of Animal Migration.* (Hodder and Stoughton, 1978)

Cox, L.B., Healey, I.N. and Moore, P.D. *Biogeography: An Ecological and Evolutionary Approach.* 3rd Ed. (Blackwell Scientific Publications, 1980)

Seddon, B. *Introduction to Biogeography.* (Duckworth, 1971)

ECOSYSTEM TYPES AND BRITISH HABITATS

Chapman, V.J. *Coastal Vegetation.* 2nd Ed. (Pergamon, 1976)

Cloudsley-Thompson, J.C. *Terrestrial Environments.* (Croom Helm, London, 1975)

Collinson, A.S. *Introduction to World Vegetation.* (George Allen and Unwin, 1977)

Cousens, J. *An Introduction to Woodland Ecology.* (Oliver and Boyd, 1974)

Eyre, S.R. *Vegetation and Soils: A World Picture.* 2nd Ed. (Edward Arnold, 1975)

Gimingham, C.H. *The Ecology of Heathlands.* (Chapman and Hall, 1972)

Hardy, A. *The Open Sea.* (Collins, 1971)

Longman, K.A. and Jenik, J. *Tropical Forest and Its Environment.* (Longman, 1974)

Pennington, W. *The History of British Vegetation.* 2nd Ed. (English University Press, 1977)

Ranwell, D. *The Ecology of Salt Marshes and Sand Dunes.* (Chapman and Hall, 1972)

Sankey, J. *Chalkland Ecology.* (Heinemann, 1966)

William, C. *Woodlands.* (Collins, 1974)

PRACTICAL TECHNIQUES

Chapman, S.B. (Ed.) *Methods in Plant Ecology.* (Blackwell Scientific Publications, 1976)

Clarke, G.R. *The Study of Soil in the Field.* 5th Ed. (Clarendon Press, 1971)

Poole, R.W. *An Introduction to Quantitative Ecology.* (McGraw-Hill, Kogakusha Ltd., 1974)

Randall, R.E. *Theories and Techniques in Vegetation Analysis.* (Oxford University Press, 1978)

Wratten, S.D. and Fry, G.L. *Field and Laboratory Exercises in Ecology.* (Edward Arnold, 1980)

ECOLOGY AND MAN

Bach, W. *Atmospheric Pollution,* (McGraw- Hill, 1972)

Bradshaw, A.D. and McNeilly, T. *Evolution and Pollution.* Studies in Biology No. 130. (Edward Arnold, 1981)

Hill, T.A. *The Biology of Weeds.* Studies in Biology No. 79. (Edward Arnold, 1977)

Irvine D. and Knights, B. *Pollution and the Use of Chemicals in Agriculture.* (Butterworths, 1974)

Leniham, J. and Fletcher, W. *Environment and Man.* Health and the Environment, Volume 3. (Blackie, 1976)

Mellanby, K. *The Biology of Pollution.* 2nd Ed. Studies in Biology No. 38. (Edward Arnold, 1980)

Open University. *Air Pollution.* Environmental Control and Public Health Units 13 and 14. (1975)

Open University. *Man and Ecology.* Ecology Units 15 and 16. (1974)

Perring, F. *The Flora of a Changing Britain.* (E.W. Classey, 1974)

Simmons, I.G. *Biogeography: Natural and Cultural.* (Edward Arnold, 1979)

Stamp, D. *Nature Conservation in Britain.* (Collins, 1969)

Steele, R.C. *Wildlife Conservation in Woodlands.* Forestry Commission Booklet 29. (1972)

Walker, C. *Environmental Pollution by Chemicals.* 2nd Ed. (Hutchinson, 1975)

Additional Background Reading

Dimbleby, G.W. *The Development of British Heathlands and Their Soils*. Oxford Forestry Memoirs. No. 23 (1962)

Elton, C.S. *The Ecology of Invasions by Animals and Plants*. (Chapman and Hall, 1958)

Good, R. *The Geography of Flowering Plants*. 3rd Ed. (Longmans, 1964)

Keith, C.B. *Wildlife's Ten-year cycle*. (University of Wisconsin Press, 1963)

Kormondy, E. (Ed.) *Readings in Ecology*. (Prentice Hall, 1965)

Odum, E.P. "Primary and Secondary Energy Flow in Relation to Ecosystem Structure". *Proceedings of the XVI International Congress, Zoology, Washington, D.C., pp. 336–338*. (1963)

Monteith, J. "Dew Facts and Fallacies". *The Water Relations of Plants*. Edited by Rutler, A.J. and Whitehead, F.H. (Blackwell Scientific Publications, London, 1963)

Phillipson, J. *Ecological Energetics*. (Edward Arnold, 1966)

Schultz, A. "The Nutrient Recovery Hypothesis for Arctic Microtine Cycles". *Grazing in Terrestrial and Marine Environments*. Edited by Crisp, P.J. British Ecological symposium, No. 4. (Blackwell Scientific Publications, 1969)

Teal, J.M. *"Energy Flow in the Salt-marsh Ecosystem of Georgia"*. *Ecology*, 43 pp. 614–624. (1962)

Walker, D. "Direction and Rate in some British Postglacial Hydroseres". *Studies in the British Isles*. Edited by Walker, D. and West, R. pp. 117–139. (Cambridge University Press, 1970)

Wynne-Edwards, V.C. *Animal Dispersion in Relation to Social Behaviour*. (Oliver and Boyd, 1962)

Examination technique

In any examination the candidate faces the problem of demonstrating his knowledge and ability to the examiner within a limited amount of time. A person's performance under these conditions can be improved if a few basic points are noted, practised and remembered.

1. Types of question. Many examination questions about ecology do not have quantitative answers. Often candidates are asked to discuss or assess various problems. However, a number of examining boards have started to include questions which are divided into many parts. Some of these ask straightforward questions to which there are quantitative answers. For example, many questions on the papers for physical geography and environmental studies have figures and maps. Typically candidates are asked a series of short questions about these. In some cases data are supplied and have to be interpreted. Despite these innovations the conventional "essay type" question remains the most frequent at "A" Level. For this reason it is useful to consider how to tackle them.

2. Understanding what is asked in the question. Many candidates lose marks because they fail to answer the question set. Instead they include much irrelevant information which does not earn them marks. It is essential that questions should be read through carefully several times and that key words should be identified. For instance, if the question asks for discussion, marks will be lost if the answer is purely descriptive. Similarly, vague generalisations and unnecessary detail do nothing more than waste time because they earn no marks.

3. Mark schemes. The examiner will have a mark scheme for awarding points exactly related to the question set. If a question has more than one section the total marks will be divided between the sections according to their relative importance. In some cases this allocation of marks will be indicated on the examination paper. If they are not, the candidate must assess these for himself. Some questions have sections which contrast markedly in importance; for example:

What do you understand by plant succession? (5 marks)
Illustrate your answer by reference to an ecological study in
which you have taken part. (15 marks)

Botany. London

Other questions have sections which carry a more balanced
distribution of marks, for example;

What is a parasite? (4 marks)
Describe the lifecycle of a named parasite. (9 marks)
Discuss the host–parasite relationship in the example you
have selected. (7 marks)

Biology. London.

4. Allocation of time. Most "A" level papers are of three hours
duration so that only 36 minutes are available for any question. It
is essential that the correct number of questions is answered and
that a reasonable amount of time is spent on each. It is pointless to
spend an hour on one question and then to rush the fourth
question or, even worse, fail to attempt it at all. The first few
marks for a question are earned relatively easily but the last few
are very difficult to acquire.

For each question the first few minutes should be spent reading
the question carefully and making a brief plan of the answer. The
construction of a plan helps to jog the memory and enables the
final answer to be presented in a concise and logical way. Time can
be saved by giving information on flow diagrams, annotated
figures, tables and simple maps. Including subheadings and
underlining key points are also useful both to candidates and
examiners.

5. Developing an examination technique. When preparing for an
examination it is very important that candidates make sure that
they really understand the work dealt with and that they revise
thoroughly. They should develop the habit of writing concisely
and clearly.

Candidates are advised to practise writing answers under the
same sort of conditions they will face at the actual examination.
When a topic has been revised, a sample question should be
looked up and then answered from memory in the time allowed.
The written answer should be checked through and an attempt
made to assess its value. In this way a candidate can obtain an
indication of his standard and progress.

Specimen questions

Figures in brackets indicate the number of marks allocated for each question.

Chapter I

1. Explain with the help of a diagram what is meant by a food web. Why is the concept of a food web more realistic than that of a food chain (11).

Plants are essential components of any food web. These harness the sun's energy in the production of organic compounds. Describe what happens to this harnessed energy in a food web (7).

(Biology "A" Cambridge)

2. An ecosystem is a cycle of materials driven by a flow of solar energy. Explain and discuss this statement (25).

(Biology "A" Cambridge)

3. Examine the energy flows and linkages within a specific ecosystem.

(Geography Special Paper London)

4. What do you understand by the term ecosystem? For one natural ecosystem show how economic development has led to its disruption.

(General Studies "A" London)

5. (*a*) Construct a simple food web to illustrate the feeding relationships you have studied in: (*i*) a named aquatic habitat (*ii*) a named terrestrial habitat (2 × 6). (*b*) For one of the above habitats describe how the feeding relationships illustrated and the behaviour of the organisms are altered by the onset of winter (8).

(Biological Studies "A, O" London)

6. Give an account of the organisms found in a named habitat. Explain how the organisms are inter-related on the basis of their nutrition.

(Biology "A" London)

7. The food web illustrated below shows some of the relationships which exist on an area of moorland. (*a*) Name the trophic levels shown in the diagram and give one example for each level. (*b*) Name one decomposer organism which is shown in the web and list two other decomposers which you might expect to

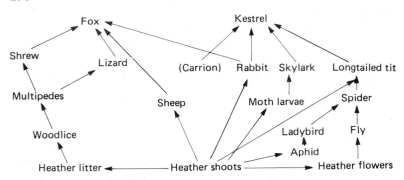

find in a moorland ecosystem. (*c*) How would you expect the numbers of organisms to alter from one trophic level to the next? (*d*) What would be the short term effects of the removal of the grazing animals from the moorland? (*e*) What would be the likely effect on the soil phosphate level of continued grazing without application of fertiliser?

(*Environmental Studies "A" London*)

8. Is the (ecological) community a well-organised system of inter-dependent species or a haphazard collection of populations in which an integration occurs by chance? Comment upon this question with reference to an area known to you.

(*Environmental Science "A" London*)

9. (*a*) What are the six main components of the ecosystem? (6) (*b*) By means of arrows show the interrelationships between these six components (7). (*c*) Cite four ways in which man can upset the balance of an ecosystem (12).

(*Geography "A" London*)

Chapter II

1. Give an account of the importance of bacteria in the natural recycling of nitrogen under the headings of (*a*) the production of nitrates (*b*) leguminous plants (*c*) denitrification (7, 7, 6).

(*Botany "A" London*)

2. The diagram illustrates the nitrogen cycle (see Fig. 8). (*a*) In what principal ways does man influence the nitrogen cycle? (*b*) Which organisms are responsible for the biological conversion of atmospheric nitrogen? (*c*) Where is the major reservoir of nitrogen located which is available for plant growth? (*d*) What is the most important effect on fresh water of increased

levels of nitrogen containing compounds? (*e*) Certain plants have the ability to fix atmospheric nitrogen. Name two such species and indicate where in the plant the process of fixation takes place.

(Environmental Studies "A" London)

3. Write an essay on "the importance to man of nutrient cycles".

(Environmental Studies "A" AEB 1980)

Chapter III

1. How would you attempt to measure production and consumption in an area of long-established grassland? To what extent are decomposers important in this habitat and how would you attempt to investigate and identify the groups of organisms involved at this trophic level?

(Environmental Studies "A" London)

2. Explain what is meant by the following terms: (*a*) (*i*) Food chain (2). (*ii*) Food webs (2). (*iii*) Pyramid of numbers (2). (*b*) Draw labelled diagrams to represent the pyramids of numbers appropriate to: (*i*) a situation in which an eagle is the final carnivore (3). (*ii*) a situation in which a protozoan parasite lives on insects in a tree (2). (*c*) Draw labelled diagrams of pyramids of energy appropriate to the two situations in (*b*) (*i*) and (*ii*) (*4*).

(Zoology "A" London, specimen for syllabus)

3. List the mineral requirements of a green plant and explain the physiological importance of any four of the required ions.

(After Biology "A" London)

4. Describe the main ways in which mammalian herbivores are adapted for the ingestion and digestion of plant material (answers may deal with more than one species) (20).

(Biology "A" Oxford and Cambridge)

5. (*a*) Define the following terms: (*i*) Biomass; (*ii*) productivity (4). (*b*) Give a named example of a species whose individuals have a large biomass but low productivity (1). (*c*) With reference to your named example explain why the productivity biomass ratio is low (2).

(Environmental Studies "A" AEB 1980)

6. Choose a food chain for which the form of the pyramid of numbers is different from the pyramid of biomass. In the space below draw the two pyramids and label them fully (4). Also draw a third pyramid showing your estimates of the approximate energy content of each link in the food chain (2).

(Zoology "A" Oxford)

Chapters IV and V

1. (*a*) With reference to either an aquatic or a terrestrial area, describe the processes which are involved in the colonisation of newly available habitats. (*b*) To what extent is a knowledge of the dynamics of ecological processes of value to the planners of new towns?

(Environmental Studies "A" London)

2. (*a*) What do you understand by the term "climatic climax vegetation"? (*b*) Name and describe briefly an important British semi-natural ecosystem that has been created by man's activities. (*c*) Indicate how this ecosystem was created and for what reason. (*d*) Describe the type of management by which it was subsequently maintained. (*e*) Outline the successful changes that would occur if human intervention ceased in this ecosystem.

(Environmental Studies "A" London)

3. "Succession" is a term which is frequently used in an ecological context. What do you understand by this term and how do successional changes take place in ecosystems? Illustrate your answer by reference to one terrestrial ecosystem and one aquatic ecosystem.

(Environmental Studies "A" London)

4. With reference to one specific example, discuss the processes and stages involved in natural plant colonization.

(Geography "A" London)

5. With reference to specific examples explain each of the following terms: plant community; plant succession; climax vegetation.

(Geography "A" London)

6. (*a*) What is meant by the term "climax plant community" (10). (*b*) List three physical factors which can arrest a plant succession (3). (*c*) Briefly explain the principal ways in which man can interfere with a plant succession (10). (*d*) Which of the following types of plant community found in the British Isles would you classify as sub-climax vegetation? (*i*) Heathland (*ii*) Fenland (*iii*) Grassland (*iv*) Oak forest (2).

(Geography "A" London)

7. Discuss the stages in the regeneration of vegetation following devastation by fire of an area of any one climax plant community. Assume there will be no interference by man.

(Geography "A" London)

8. Illustrating your answer with reference to specific examples,

explain how and why plant communities change with time.

(Geography "A" London)

9. What do you understand by plant succession? Illustrate your answer by reference to an ecological study in which you have taken part (5, 15).

(Botany "A" London)

Chapters VI and VII

1. Define transpiration (3). Give an account of the ways in which air and soil conditions affect the rate of transpiration by a plant (17).

(Biology "A" Oxford and Cambridge)

2. From your own field observations describe the range of physical and biotic factors which may affect the distribution of animals.

(Zoology "A" London)

3. The sketch shows a vertical profile through a rendzina soil. (*a*) With what types of parent material are such soils associated? (*b*) Name two areas of the British Isles where you would expect to find rendzina soils. (*c*) What kind of vegetation would you expect to find in these areas? (*d*) Why is the topmost horizon relatively shallow? (*e*) Why were such areas important for neolithic people in Britain?

(Environmental Studies "A" London)

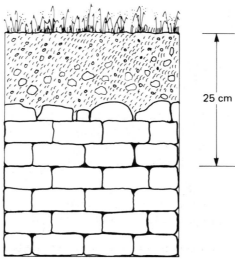

25 cm

4. Describe the climatic, edaphic and biotic factors which act on an area of land recently devastated by fire. Describe the changes in the vegetation which would take place in the burnt area if it were left uncultivated (10, 10).

(Botany "A" London)

5. (*a*) Describe the climatic, edaphic and biotic factors which act on newly made banks and cuttings along major roadways. (*b*) What measures are taken to ensure a covering of plants on these areas and why should such a covering be encouraged? What is the significance of these banks and cuttings to conservation? (10).

(Botany "A" London)

6. Light is as important for animals as it is for plants. Discuss this statement (25).

(Biology Special Paper London)

7. Define the terms plant community and biotic factor. Explain and discuss the changes which may take place as a result of biotic factors operating either over a period of one year or a period of several years in any one community you have studied (5, 15).

(Botany "A" London)

8. What is the distribution of stomata in the following: (*a*) the floating leaves of water plants (*b*) submerged leaves of water plants (*c*) xerophytic leaves (*d*) mesophytic, dorsiventral leaves (*e*) grass leaves (total 5)

(Botany "A" London)

9. Trace the passage of water from the vacuoles of leaf cells to the external atmosphere. How may this movement be affected by (*a*) the structural features of the leaf and (*b*) the external atmosphere? (6, 7, 7).

(Botany "A" London)

10. You are provided with a monolith (undisturbed soil section) taken from a podsolised brown-earth soil. (*a*) What mineral constituents would you expect to find in the "A" horizon of the soil? (*b*) There is an eluvial horizon present. What is meant by this term? (*c*) What is the general name given to organisms which assist in the breakdown and incorporation of dead waste materials in the soil? Give the names of three such organisms. (*d*) What particular physical characteristics of humus, when incorporated in this soil, improve the soil's structure and moisture holding capacity? (*e*) Why is this effect not found in the podsols?

(Environmental Studies "A" London)

11. Explain what is meant by the "moisture budget" of a soil. Discuss the factors that affect it.

(Geography "A" London)

12. List the types of ecological factor that affect the life of terrestrial animals. How does each type of factor affect terrestrial animals in general? Show how one factor of each type affects the life of an individual of a named species.

(*Zoology "A" Oxford*)

13. (*a*) Compare, with the aid of annotated diagrams, the profile characteristics of two distinct soil types which you have studied (8). (*b*) Discuss the factors which caused the development of each of these two soil types (8). (*c*) (*i*) Define the term "climax vegetation" (2); (*ii*) Specify one of the soil profiles. already discussed and describe a climax vegetation associated with it (2).

(*Environmental Studies "A" AEB 1980*)

Chapter VIII

1. Describe the evolutionary history of one named major group of animals. (17). What general principles does this particular sequence illustrate? (8).

(*Zoology "A" London*)

2. What do you understand by the term mutation? Outline the importance of mutations in plant speciation and plant evolution (5, 15).

(*Botany "A" London*)

3. Define the term species (7). Survey the ways in which new species might arise, giving one example of each mechanism you describe (18).

(*Zoology "A" London*)

4. Give an account of the origin of new varieties and species of plants (30).

(*Botany "A" London*)

5. Give an account of the different sources of variation that occur in plants and animals. Illustrate your answer by reference to variations you have observed personally in wild species.

(*Biology "A" London*)

6. Present a selection of the evidence which you think might convince a sceptical friend that evolution has taken place.

(*Biology "A" London*)

7. "Evolution is still occurring." Present evidence in favour of this view.

(*After Biology "A" Oxford and Cambridge*)

8. In industrial areas of England it has been reported that the dark form of a certain species of moth has increased in numbers compared with the lighter form. Explain how such a change may

have come about. How would you attempt to demonstrate the truth of your explanation?

(Biology "A" London)

9. What is the evolutionary significance of each of the following? (*a*) Industrial melanism in moths? (3); (*b*) Resistance of insect pests to insecticides (3); (*c*) Geographical isolation (3); (*d*) Fossil remains (3).

(Zoology "A" Univeristy London, specimen for syllabus)

10. "The fact of evolution by natural selection as proposed by Charles Darwin is now universally accepted by all competent to express an opinion, and its mechanism has been, in principle, explained." Critically discuss this statement made recently by a distinguished biologist.

(Biology "A" Oxford)

11. Define as concisely as possible the following in terms of the Darwinian theory of evolution: (*a*) the struggle for existence (2); (*b*) Natural selection (2); (*c*) the origin of species (2).

(After Zoology "A" Oxford)

Chapters IX, X and XI

1. (*a*) In a short account of a *named* internal parasite (e.g. a platyhelminth or a nematode) of a vertebrate animal, describe *i*) its method of nutrition, and *ii*) any adaptations to parasitism. (*b*) Many parasites spend some periods of their lifecycles living independently of their hosts. Discuss briefly the advantages and disadvantages to the parasite of having an independent stage in its lifecycle.

(Biology "A" Cambridge)

2. What is a parasite? Describe the lifecycle of a named parasite. Discuss the host–parasite relationships in the examples you have selected (4, 9, 7).

(Biology "A" London)

3. (*a*) What do you understand by symbiosis? (*b*) Discuss symbiosis in *i*) Lichens and *ii*) Root nodule bacteria (5, 7, 8).

(Botany "A" London)

Chapter XII

1. Examine the similarities and differences between a temperate deciduous forest ecosystem and a temperate coniferous forest ecosystem.

(Geography "A" London)

2. Examine the similarities and differences between a tropical grassland ecosystem and a temperate grassland ecosytem.

(Geography "A" London)

3. Either: (*a*) Discuss the effect of man on the ecosystem of the tropical rain-forest; or (*b*) describe the characteristics of the moorland and heathland ecosystems in Britain and discuss the problems of their conservation.

(Geography "A" Oxford)

4. Why are similar climatic zones not always associated with similar assemblages of plants?

(Geography "A" Oxford)

Chapter XIII

1. What do you understand by the term "sampling" in an ecological investigation? Describe briefly an investigation in which you have used a sampling technique and the manner in which sampling was carried out. Name and describe a method of testing the significance of your results. How might the sampling technique influence your results?

(Environmental Studies "A" London)

2. (*a*) Define the following terms, so as to bring out their differences in meaning, as used in ecology. (*i*) Community (*ii*) Habitat (*iii*) Ecosystem (*b*) Explain briefly how you would use each of these terms in the context of an area you have studied in detail.

(Environmental studies "A" London)

3. (*a*) (*i*) Give an account of the techniques you would employ to measure the distribution of organisms in a salt marsh (7) (*ii*) What results would you expect to obtain from such a study? (7) (*b*) Explain why sandy and muddy shores, though also having soft sedimentary substrates are not readily colonised by plants.

(Biological Studies "A, O" London)

4. Describe the major ecological features of a specific habitat which you have studied in the field (15). Construct a food web showing the possible interrelationships of the organisms occurring in this habitat (10).

(Zoology "A" London)

5. You have been asked to investigate the fauna of a specific habitat. By reference to a named site, of which you have personal field experience, show how you would set about carrying out such a survey in both qualitative and quantitative terms (15). In what ways do you think the results of such an investigation might be

beneficial both to fellow zoologists and to other users of this habitat (10).

(Zoology "A" London)

6. You have been asked to prepare a detailed ecological survey of an area near your home which has been notified as an SSSI. The area is 4 ha. in extent and has scrub and grassland habitats in approximately equal proportions in undulating terrain. Describe how you would carry out such a survey and comment upon the reliability and accuracy of your methods in providing a full report of the features of the site (1 ha = 2.47 acres).

(Environmental Studies "A" London)

7. Describe in detail how you would carry out a field experiment designed to investigate one of the following topics on a hill or mountain slope having a vertical range of approximately 500 metres: (*i*) The relationship between slope form and geology; (*ii*) The relationship between altitude and soil characteristics; (*iii*) The relationship between soil characteristics and degree of slope; (*iv*) The relationship between altitude and vegetation cover.

(You may refer, if you wish, to statistical tests that might appropriately be used to analyse data collected in the field.)

(Environmental Studies "A" London)

8. Answer the following question by reference to one named plant community you have studied. (*a*) Name of community (1) (*b*) (*i*) State the dominant plant (*ii*) State how you obtained this information (2) (*c*) Name four other plants growing in this community (2) (*d*) Briefly describe the character of the soil type of this area (3) (*e*) Give an account of one quantitative method you have used during the course of your work on this community (4).

(Botany "A" London)

9. What edaphic and biotic factors determine the diversity, distribution and numbers of species to be found in a habitat? Illustrate your answer by reference to an area which you have studied mentioning how you estimated the diversity, distribution and numbers of particular named plants and animals.

(Biology "A" Oxford)

10. With reference to techniques used in ecological studies with which you are familiar discuss their limitations and ways in which they may be improved.

(Environmental Studies "A" AEB 1980)

11. From a large isolated population in Dorset, 100 peppered

moths, *Biston betularia*, were captured. They were marked with a spot of cellulose paint and then released. Later 1000 individuals were captured from the same population in a light trap and of these only ten have the paint spot. What is the size of the population?

(After Zoology "A" Oxford)

12. Making reference to an area which you have studied describe the techniques used in carrying out investigations to discover: (*a*) general trends in biotic gradients including distribution of organisms; (*b*) environmental factors which may influence biotic distribution; (*c*) density of an individual species; (*d*) biomass estimations (20).

(Environmental Studies "A" AEB 1979)

Chapter XIV

1. "Man is destroying his environment." Discuss this statement.

(Biology Special paper, London)

2. What are the ecological and environmental implications of the need to increase world food production?

(Environmental Studies "A" London)

3. Examine carefully the kind of environmental damage that has occurred from river pollution. What action can be taken to improve the situation?

(General Studies "A" London)

4. (*a*) Indicate briefly the main effects on wild life in Britain of the following trends in agricultural practice (*i*) Increase in average field size (*ii*) Increased use of fertilizers (*iii*) Increased use of herbicides and insecticides (*iv*) Improved techniques of land drainage. (*b*) Briefly describe two possible ways in which the impact of modern agriculture on wild life may be reduced.

(Environmental Studies "A" London)

5. Discuss the conservation of biotic resources.

(Geography Special paper, London)

6. (*a*) Discuss the various causes of river pollution. (*b*) Examine the consequences of pollution to biotic balances within a river system.

(Geography "A" London)

7. Explain the significance of vegetation in soil conservation.

(Geography "A" London)

8. Conservation may be regarded as an important botanical activity. Discuss this statement (30).

(Botany "A" London)

9. The diagram illustrates the percentage saturation of dissolved oxygen in the River Thames above and below London Bridge during the months of July, August and September in 1963, 1973 and 1975.

(*a*) Why, on all three curves, do oxygen levels fall as London Bridge is approached from upstream? (*b*) What biological changes would you expect to accompany a fall in oxygen levels? (*c*) What factors explain the general increase in oxygen levels between 1963 and 1975? (*d*) Surface-feeding duck, which feed largely on Tubifex worms, became common below London Bridge in the late 1960s; by 1975 the numbers of these birds were declining. Can you suggest an explanation for this? (*e*) The levels of dissolved oxygen in 1976 indicated a deterioration from the situation compared to 1975. Suggest a reason for this.

(Environmental Studies "A" London)

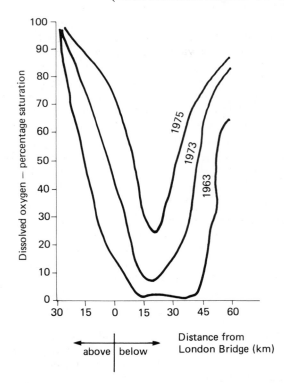

(From C.J. Krebs, Ecology, *Harper and Row, 1978)*

10. In what ways does the knowledge of plants and plant processes contribute to modern practices in either medicine or forestry? (20).

(Botany "A" London)

11. (*a*) In what ways has man brought ecological benefit to other organisms (*i*) Intentionally; (*ii*) Inadvertently? (*b*) In what ways has he increased the ecological problems of other organisms (*i*) Intentionally (*ii*) Inadvertently? (*c*) Discuss the justification for the intentional actions referred to in (*a*) and (*b*). (*d*) Discuss the feasibility of reducing the harmful effects to other organisms of his actions.

(Zoology "A" Oxford)

12. In what respects can present day agricultural and forestry practices be said to (*a*) Improve (*b*) be detrimental to, our natural environment? Illustrate your answer by means of relevant examples with which you are personally familiar.

(Biology "A" Oxford)

13. In recent years increasing resources have been made available for the preservation of endangered species such as the golden eagle, the osprey, the otter and the Snowdon lily. It could be argued that with limited resources at our disposal, time and energy should be better spent on more urgent environmental problems to the consequent detriment of endangered species. Compose a balanced discussion on the basis of these points of view.

(Environmental Studies "A" AEB 1979)

14. Read the following paragraph and answer the questions which follow.

"Use of energy is the principal source of air pollution. Energy production, transportation and consumption are responsible for an important fraction of all our environmental problems. Use of energy continues to rise at the rate of 2% per annum. Even if fuel supplies were infinite, such an increase could not be tolerated indefinitely. But fuel supplies are not inexhaustible, and this, combined with the need to preserve the environment, will force changes in patterns of energy production and use."

(*a*) Outline the main kinds of air pollution caused by the use of energy (3). (*b*) (*i*) What legislation has been introduced to ameliorate air pollution in the United Kingdom? (1) (*ii*) How effective has this legislation been? (2) (*c*) What other environmental problems are the result of energy production, transportation and consumption? (4). (*d*) Explain briefly why energy use has risen at the average rate of 2% per annum (2). (*e*) Suggest, with reasons, what changes in the United Kingdom's patterns of energy production and use may occur during the next fifty years (8, 20).

(*Environmental Studies "A" AEB 1979*)

Index